Women of Genius in Science

Lars Jaeger

Women of Genius in Science

Whose Frequently Overlooked Contributions Changed the World

 Springer

Lars Jaeger
Baar, Zug, Switzerland

ISBN 978-3-031-23925-0   ISBN 978-3-031-23926-7   (eBook)
https://doi.org/10.1007/978-3-031-23926-7

This Springer imprint is published by the registered company Springer Nature Switzerland AG
The registered company address is: Gewerbestrasse 11, 6330 Cham, Switzerland

To my daughter Kira Anh Jaeger

Foreword

This is one of the most comprehensive books on the history of women in science I've ever read. In a timeframe spanning nearly seventeen centuries, since the year 355 to current times, Lars Jaeger eloquently describes the struggles of eighteen brilliant female scientists in the fields of philosophy, physics, medical sciences, mathematics, astronomy, astrophysics, computer science, chemistry, and primatology.

In this book, we can read how, in the early fifth century, Hypatia of Alexandria was fascinated with the ellipse, applying it to explain planetary orbits in a heliocentric worldview (eight centuries before Kepler!) and how her non-dogmatic critical thinking led to a violent outcome, marking the turning point in a dogmatic anti-educational and anti-scientific Christian era that lasted over a millennium. A big leap in time takes us to the work in the Middle Ages of Hildegard von Bingen and her notes on observational medicine, human physiology, feminine sexuality, and the healing power of plants, herbs, and minerals. Although based on a deep religious faith, her studies helped lay the foundations of the scientific method, established 600 years later, when Newton came up with the laws of mechanics. This is when Émilie du Châtelet gave an important impulse to the early Enlightenment by changing the mathematical structure of Newton's Principia and translating it from Latin to French, making it comprehensible to a wider audience in Europe. We can read here how she depended on her strong determination and self-esteem to get her papers through to the Académie des Sciences in

Paris and how her conclusions on the nature of light were also 170 years ahead of her time.

In the late eighteenth century, the Enlightenment spread to Italy mainly thanks to the scientific work on electricity and magnetism by the physicist Laura Bassi, the first woman to be elected to the Bologna Academy of Sciences (or to any other academy or university), even though she had never been able to receive a scientific training at a public university herself. Like other curious and talented women, Sophie Germain also had to strive to get a proper scientific education (mathematical in her case) and she also profited from the support of her family. Lars Jaeger takes us through her outstanding scientific life and explains how she also had to hide behind a male identity to break her way through. Sophie Germain then is an example of a largely forgotten scientist to whom humanity owes proper recognition: In spite of her breakthroughs in mathematics, her name does not appear on a list of exceptional scientists and engineers of the eighteenth and nineteenth centuries in a prominent place in Paris. And, as you might have guessed, no other woman is included on that list.

The following chapters delve into the lives, achievements, and struggles of other prominent scientists like Caroline Herschel, astronomer, one of the few scientists of the nineteenth century whose work was appreciated during her lifetime and the first to receive a salary for her work; Ada Lovelace, the pioneer of modern computer science, dubbed "the first computer programmer in history," or "the inventor of software," her ideas and concepts preceding modern computer science developments by at least a century; the Russian mathematician Sofia Kovalevskaya, the first female professor of mathematics to lecture independently (at the Swedish University in Stockholm) and considered the most important Russian mathematician of the nineteenth century

Of course, Marie Skłodowska Curie is the great and most famous female physicist whom I took as a role model during my own career. She was not allowed to present her revolutionary research on radioactivity at the Académie des Sciences in Paris in 1898. It had to be presented by her doctoral supervisor, because women were not allowed to be members of the Académie. However, she was appointed as the first female faculty member of the École Normale Supérieure in 1900. Her life and struggles are extremely well grasped and exposed in this chapter. Lise Meitner, on the contrary, did not find proper recognition by the public in spite of being the discoverer of nuclear fission, in close collaboration with Otto Hahn, thus paving the way to nuclear energy. It is hard to believe that, less than a century ago, she had to put up with such poor conditions for her research just because she was a woman, and that in spite of that she made such important discoveries!

Emmy Noether deserves a separate chapter. A German mathematician in the early twentieth century, she found a beautiful and powerful connection between the symmetries of the basic mathematical equations of physics and the conservation laws of nature. She, too, had to struggle past significant obstacles to make her career. How difficult it must have been in 1915 if among the faculty protocols at the University of Göttingen, one could read (regarding her habilitation): "Are we not of the opinion that a female head can only be creative in mathematics as a very exceptional case?" I must confess that when I learned "Noether's theorem" during my bachelor's degree in physics, I did not know she was a woman, and only learnt this several years later! And her fame as "the mother of modern algebra" is well deserved. However, she never obtained a standard professorship at a university.

Even in the twentieth century, Jaeger describes the lives of outstanding female scientists in which we only see timid and totally insufficient improvements in the recognition of women. For example, the German mathematician Grete Hermann identified a basic mistake in the fundamental interpretation of quantum physics in the 1930s, by appeal to her deep philosophical insight. These and subsequent developments were only proven several decades later by John Clauser, Alain Aspect, and Anton Zeilinger, who were awarded the very recent 2022 Nobel Prize in Physics for that.

The following chapters recount the extraordinary lives of other female scientists that belong to contemporary science. One is the Chinese physicist Chien-Shiung Wu, who was also discriminated against for her different culture as a migrant in the USA. Her involvement in the Manhattan Project is described in detail, along with her important discoveries in nuclear and particle physics. But, once again, as happened with other exceptional female physicists, she was excluded from the Nobel Prize. Rosalind Franklin is now well known for having been left out of the Nobel awards, even though her fundamental role in the discovery of the structure of DNA is well established. Details of this unfair process and on how her discoveries were underestimated are well described in the chapter. I really enjoyed the chapter on the achievements of the British primatologist Jane Goodall, who is an example to the younger generation on conservation and care of the environment and who overcame several difficulties with passion.

It is interesting to compare the contemporary astrophysicist Jocelyn Bell-Burnell, the discoverer of radio pulsars, with Caroline Herschel because it gives us an idea of how conditions have changed for women in science after more than two centuries. I leave this exercise to the reader, but one can see that, once again, the highest award was kept away from a female scientist who fully deserved it. However, Jocelyn Bell-Burnell later received other important awards for her exceptional discoveries.

The book ends with a fair tribute to two other outstanding scientists: Lisa Randall, the first woman to be appointed as chair of theoretical physics in Princeton (as late as in 1998!), who made important contributions to the understanding of cosmological inflation, dark matter, and string theory. The last chapter is devoted to Maryam Mirzakhani, a brilliant Iranian mathematician, the first woman to be awarded the Fields Medal in 2014, the highest award in mathematics, for her work on complex geometry in abstract spaces. She once said that there remain barriers for girls interested in mathematics and that balancing work and family is still a big challenge.

Maryam Mirzakhani was right: Although the conditions have improved and more women are following scientific careers, the situation is far from ideal. In this book, Lars Jaeger walks us through the history of science focusing on the lives of eighteen exceptional women who have had to overcome too many difficulties, impositions, and discriminations. These women were strong and they fought. I wonder how many women have been left behind. May this book serve as an inspiration to future generations.

Prof. Dr. Karen Hallberg
Centro Atomico Bariloche
San Carlos de Bariloche, Argentina

Contents

1

Hypatia of Alexandria (ca. 355–415 or 416): Icon of Mathematics in Late Antiquity

For many centuries, Alexandria in northern Egypt was the city of knowledge. Its foundation by Alexander the Great in 332 BC coincided with a time when Plato and Aristotle were launching their Athenian schools of thought and people were excited by intellectual diversity and discursive science. The city, which had become rich through trade, would soon afford itself with the most important collection of books in the ancient world. The Greek historian Strabo reports that Aristotle was personally involved in the planning of the library and determined the way in which knowledge should be organised in it. Within a few decades, 400,000 to 500,000 papyrus scrolls were collected, and some sources even speak of 700,000 or even more manuscripts.

The library became a meeting place for scholars from all parts of the known world. Many of those scholars of antiquity whom we still remember worked at least temporarily in Alexandria, including the physician Herophilus of Chalcedon, the engineer and mathematician Heron of Alexandria, and the astronomer Aristarchus of Samos. The two most important mathematicians of antiquity are also inextricably linked to this city: Archimedes of Syracuse and the Alexandrian Euclid. Even after the conquest of Alexandria by the Romans in 30 BC, the city acted as a magnet for physicians, philosophers, philologists, mathematicians, geographers, and astronomers. When discussing science, historians speak of the period between about 300 BC and 300 AD not as Hellenistic or Roman, but as Alexandrian, because it was here in north Africa—not in Athens or Rome—that scientific thinking first blossomed.

© The Author(s), under exclusive license to Springer Nature
Switzerland AG 2023
L. Jaeger, *Women of Genius in Science*,
https://doi.org/10.1007/978-3-031-23926-7_1

But from about 300 AD, Alexandrian culture went into decline. The library dwindled to insignificance, and within a few decades a large part of the ancient knowledge was irretrievably lost. Around 400 AD, this dream of an undogmatic pursuit of knowledge was over. How could it have come to this? The Hellenistic tradition of thought, which was characterised by curiosity and tolerance, had survived its integration into the culture of the Roman conquerors almost unscathed. But now a new spiritual orientation took its place: Christianity. It focused more on the afterlife than on this world, and interest in a scientific understanding of the world disappeared. The Council of Nicaea, convened by the Roman Emperor Constantine I, in which Christianity was elevated to the status of a state religion, took place in 325 AD. At first, other religions were still allowed to practise, but 55 years later, this tolerance was finally quashed: Emperor Theodosius declared Christianity the *sole* state religion in 380 AD.

With the triumph of Christianity, a manageable number of dogmas took the place of diverse knowledge and lively exchange. Now there was no longer any interest in maintaining and caring for extensive libraries. The gigantic storehouse of knowledge and literature in Alexandria suffered the same fate as similar institutions everywhere in Europe, the Near East, and North Africa. From around 390 AD until the fall of the Roman Empire in 476 AD—the beginning of the early Middle Ages—the stock of titles fell from more than a million to a maximum of one thousand. In other words, only one in a thousand books was deemed valuable enough by Christianity to be regularly copied and saved from oblivion.

This loss of knowledge set the learned world back fifty generations. It was not until over a thousand years later that a rudimentary scientific way of thinking re-emerged in Europe, and another two centuries went by before the level of knowledge of antiquity was reached again. For example, Eratosthenes had determined the circumference of the Earth with astonishing precision in Alexandria in around 230 BC; depending on the conversion factor from Greek stadia to present-day kilometres, he had achieved an accuracy of up to 99 percent. But his work was lost; until the sixteenth century, calculations were made using a value that was 10,000 kms too small. It had been calculated in 150 AD by Ptolemy, who was highly revered and much copied in Christianity—this was the only reason why Columbus believed he had discovered India in 1492.

* * *

Many books that offended the new and powerful church crumbled to dust. Others were deliberately destroyed, as the spiritual and intellectual tolerance associated with religious diversity also gradually eroded. After only eleven years of Christianity as the state religion, all pagan temples in the Roman Empire were closed by law; all non-Christian books that could be seized were probably burned on this occasion. A further wave of destruction was triggered by the law enacted by Theodosius' successor Honorius in 408 AD:

> If any images are still standing in temples or shrines, and if they received worship from pagans anywhere today or ever before, let them be pulled down.

In the following year, 409 AD, another imperial law obliged all mathematicians to burn their books before the eyes of their bishops. Otherwise they were to be expelled from the Roman Empire. Things were getting more difficult for the followers of ancient culture and knowledge; the possession of books that met with the disapproval of the Church could cost them their lives. Even the tiniest of niches in which the remnants of Alexandrian traditions of thought still survived were gradually cleared out. Only Arab culture profited from the exodus of scholars who did not want to come under the umbrella of the Church, despite its dogma of being the only one that could bring salvation. The fact that Arab culture was able to undergo such an unparalleled expansion three hundred years later was due not least to the science and philosophy of the refugees.

* * *

In this time of upheaval, the mathematician, astronomer, and philosopher Hypatia lived in Alexandria. In an environment already dominated by Christianity in the second half of the fourth century, she probably belonged to a non-Christian minority. It is certain, however, that she was murdered in 415 or 416. Unfortunately, after her death, evidence of her work was so extensively erased that only secondary sources allow conclusions to be drawn about herself and her thinking.

- The letters of Synesius of Cyrene to Hypatia are probably the most productive sources; he also mentions her in numerous other letters to other people. Synesius made a career as a Christian and became bishop of Ptolemais, today's Acre in Israel. As a student and friend of Hypatia, however, he had also internalised impartiality and tolerance. His written testimonies are therefore characterised by an astonishing non-partisanship. In his short treatise "On the Gift" (*Pros Paiónion perí tou dõrou*), he presents Hypatia as his "most venerable teacher". In a letter written around 395 to his good friend Herculianus, Synesius describes Hypatia as

(...) such a famous person, her reputation seemed literally unbelievable.[1]

Again and again he praises her philosophical steadfastness in challenging times.

- Socrates Scholasticus was also a contemporary of Hypatia; whether he was a Christian or not is unknown. In his main work, the seven-book ecclesiastical history *Historia ecclesiastica,* he writes:

In Alexandria lived a woman named Hypatia (...) She had such an outstanding education that she outshone all the philosophers of her time. Her teaching brought her to the top of the Platonic school (...) To the authorities she appeared frankly and with the self-confidence that her education gave her, and she also showed no shyness about being in the company of men. Indeed, because of her exceptional intelligence and strength of character, everyone met her with awe and admiration.[2]

He explicitly states that Hypatia belonged to the school that Plotinus had founded and that represented the predominant Neoplatonic philosophy. Socrates emphatically condemned the murder of Hypatia as an unchristian act:

Surely nothing can be further from the spirit of Christianity than the approval of massacres, fighting and transactions of this kind.[3]

- An epigram by the Alexandrian poet Palladas, written during Hypatia's lifetime, reads:

When I see you, hear your word, I adore you, beholding the star-covered house of the noble virgin; for to the heavens alone extend all your doings, your ornament and adornment of every speech, Hypatia, the pure, unsullied star of the highest wisdom![4]

---

[1] Joseph Vogt, *Begegnung mit Synesios, dem Philosophen, Priester und Feldherrn. Collected Contributions.* Darmstadt (1985).

[2] Günther Christian Hansen (ed.), *Sokrates Kirchengeschichte - Band 1:Die griechischen christlichen Schriftsteller der ersten Jahrhunderte.* Berlin: Akademie-Verlag (1995).

[3] Ralph Novak, *Christianity and the Roman Empire.* Harrisburg, PA: Trinity Press International (2001).

[4] Georg Grützmacher, *Synesios von Cyrene: ein Charakterbild aus dem Untergang des Hellenentums.* Leipzig: A. Deichert'sche Verlagsbuchhandlung (1913), p. 23.

- The Neoplatonist Damascius (born 480, died after 538) was the last head of the Neoplatonist school in Alexandria. His surviving but fragmentary works "Philosophical History" and "The Life of the Philosopher Isidoros", written about a hundred years after Hypatia's death, contain some statements about her life, including:

> The whole city rightly loved her and revered her remarkably, but the rulers of the city envied her from the beginning.[5]
>
> He also writes that she possessed a greater genius than her father, who was her first teacher.

- In the "Suda", a Byzantine encyclopaedia of the tenth century, an entire article is dedicated to Hypatia.[6] It is a juxtaposition of various sources, including that of Damascius. Five hundred years after Hypatia's death, events were for the most part presented in such a tendentious manner that the truthfulness of the articles becomes questionable. The human being Hypatia had become a legend.

* * *

What picture emerges of the legendary Hypatia? Almost all sources mention that she was an outstanding mathematician and philosopher. She was trained in mathematics and astronomy by her father Theon of Alexandria. The latter was himself a famous scholar. In his new edition of Euclid's Elements, he corrected all the spelling and transcription errors that had crept in over the course of almost 700 years of uninterrupted copying. This edition of the "Elements" almost completely displaced all other versions and is the most widely used edition of Euclid's textbook of mathematics to this day—at the same time, it was the last scientific work by a known author to be included in the Library of Alexandria.

It is not known who Hypatia's philosophy teacher was. She probably attended the Neoplatonic school in Alexandria. However, it is also reported that she travelled around the city wearing a philosopher's cloak, a garment adopted in particular by the followers of the Cynic philosophy. Presumably she taught a philosophy which was based on Neoplatonism, but which she enriched with Cynic ideas. Cynics were a challenge to society, because

---

[5] Johann Rudolf Asmus (ed.), *Das Leben des Philophen Isidoros von Damaskios aus Damaskos*. Leipzig: Meiner (1911). The original of the historian Damascius is lost.

[6] Ibid.

they resisted social constraints in a confrontational way and questioned the way of life of others through their rejection of material comforts. In addition, Hypatia remained unmarried her whole life, and virginal, according to almost all traditions; at the same time, various sources mention her extraordinary beauty—Damascius describes Hypatia as extraordinarily handsome and beautiful in figure. It is therefore likely that her appearance and behaviour were interpreted as shameless and disturbing by people who were not among her followers. This brought her into conflict, especially with the Christians, who already strongly rejected criticism and provocation in the early phase of their power. Hypatia's call for tolerance did not reach them, because in their view the world was built according to God's rules, which were not to be questioned.

Hypatia also had enemies among the philosophers of Alexandria. For those who saw themselves as the intellectual elite, it was an imposition that she did not discuss philosophical topics in learned circles, but preferred to teach in the streets of Alexandria. With charisma and charm, she interpreted the famous teachings of Plato, Aristotle, and every other philosopher; to any mathematical questions she provided answers and made complicated concepts understandable to her listeners. Several sources describe her lectures as excellent and she herself as extremely popular with her listeners. She had no official teaching position for which she was remunerated, but probably made her living by collecting contributions directly from her audience.

* * *

Hypatia's declared goal was to preserve, explain, and pursue the philosophical and mathematical heritage of Alexandria. In mathematics, Hypatia was primarily concerned with the works of Euclid and Archimedes. Unfortunately, no concrete mathematical statement has survived that can be attributed to Hypatia with absolute certainty, but numerous sources attest to her numerous insights, with great significance for the mathematics of late antiquity. An example of the fact that some of her works probably only survived through the ages *because* her name was not mentioned is the commentary on Ptolemy's main work, the *Almagest,* written in the second century AD—the name of this summary comes from the Arabic term "al-maǧistī," which means "the great synthesis".

Hypatia's father Theon also revised this standard work of antiquity, which summarises all astronomical knowledge. In the oldest edition of the commentary he wrote, he notes in the heading to the third book that it was a version "reviewed by the philosopher Hypatia, my daughter". For a long time, it was assumed that Hypatia had only reviewed her father's corrections. Today, however, it is assumed that Hypatia did not just correct her father's commentary, but actually corrected the text of the *Almagest* itself. In addition, she improved the method for the long division algorithms needed for astronomical calculations. Some historians even assume that Hypatia edited and re-edited not only Book III, but all nine extant books of the *Almagest*.

The Byzantine Suda attributes other significant mathematical works to Hypatia:

- An important commentary on the arithmetic and algebra of Diophantus of Alexandria: Diophantus, who presumably lived more than a hundred years before Hypatia, is widely regarded today as the "father of algebra." He dealt extensively with linear and quadratic equations, and in some cases even cubic equations, in his thirteen-volume magnum opus *Arithmetica*. Presumably, the books of Diophantus Hypatias, which have been copied again and again over the centuries, contain extensive commentaries which at some point were judged to be the original text. Experts think this is likely because the Arabic books are considerably more extensive compared to Diaphantus' original book. (Of course, the Arabic books also contain new ideas in the field of algebra. Indeed, the word "algebra" also comes from the title *cIlm al-jabr wa l-muqābala* (The Science of Restoring and Balancing) of a book by the great Persian mathematician al-Khwarizmi.) Historical research of the nineteenth century cites Hypatia as one of the essential sources of the additional material that the Arabs had at their disposal, compared to the original text by Diophantus. Apart from her, no other mathematician is known to have written a commentary on the *Arithmetica*. Moreover, the additions show her methodical way of writing.
- An important commentary on Apollonius of Perga's work on conic sections. According to tradition, Hypatia was fascinated by the ellipse, a figure that results when a plane is made to cut through a cone. Twelve hundred years before Johannes Kepler, she tried to explain the irregular orbits of the planets in terms of this geometric figure.
- She is also credited with the lost text "On the Astronomical Canon", which provides a mathematical description of planetary motion. It is highly likely that this was a new edition of Ptolemy's astronomical tables or the already mentioned commentary on his *Almagest*. In this work, Hypatia worked with Ptolemy's geocentric world view, the only accepted doctrine in the Occident, and indeed in the Orient, until the sixteenth century, but also with the heliocentric world view put forward by the Greek mathematician Aristarchus.

- The anonymous text "Measuring the Circle" on isometric figures, in which, among other things, the number π is delimited very precisely, is usually attributed to Archimedes. But it may have been Hypatia who published this book more than two hundred years later.

All these achievements require a very high level of mathematical knowledge and even genius. Most scholars who study this subject now recognise that Hypatia must have been one of the leading mathematicians, not only of antiquity, but of all time. This view cannot be proven beyond doubt, but it seems highly probable.

The correspondence with the already mentioned Synesius of Cyrene shows that Hypatia was also interested in mechanics. Numerous drawings of a variety of instruments have survived, including an astrolabe. With this apparatus, the date and time can be determined on the basis of the positions of stars and planets, and the positions of the celestial bodies at any given date in the future can also be read. Although devices of this kind had already been known for at least 500 years in Hypatia's time, a further development making use of two rotating disks may have originated from her; this facilitates the tasks of spherical astronomy. In another letter, Synesius asks Hypatia to build him a hydroscope. The instrument, known today as a hydrometer, determines the density or specific gravity of liquids. Such instruments already existed in the second century BC, but Hypatia may well have improved this apparatus with her own ideas.

* * *

In the early fifth century, when Hypatia was about 60 years old, the tension between free-thinking philosophers and followers of the Christian religion was heading for a climax. It was much more than a conflict between people striving for knowledge and those who thought they already knew everything through God. It was above all about political power.

From October 412, Cyril ruled as Christian patriarch in Alexandria. His comparatively tolerant predecessor Theophilus (and uncle, Cyril was the son of Theophilus' sister) was even a follower of Hypatia's school, although he had had non-Christian temples destroyed, But Cyril was a hardliner, and systematically incited hostilities against Jews as well as philosophers and scientists who had not rallied the orthodox teachings of the Christian Church. His opponent was Orestes, the Roman governor in Egypt, and thus the highest-ranking state representative, whose task was to maintain public order. The Christian Orestes was a pupil of Hypatia and on friendly terms with her. Thanks to his cosmopolitan attitude, he moved quite naturally in non-Christian circles. He struggled to fend off Cyril's constant attempts to interfere in the political order and thus in worldly affairs.

Hypatia got caught up in the bitter power struggle between Cyril and Orestes, i.e., between spiritual and state power. As far as we know, she made no statements against Christianity, but this educated, free-thinking, and independently-acting woman, who moved undaunted in public circles, even among men, and who carried out "pagan propaganda" with her philosophy and science, was viewed as an intolerable provocation by the fanatic Cyril. In addition, she was popular among the people of Alexandria and possessed political clout, since Orestes took advice from her in political and legal matters. As a follower of scientific rationalism, she very likely refused to give up her ideals and become a Christian, and this may have contributed to her demise.

Cyril targeted Hypatia in order to harm Orestes, who remained out of his reach. In his work "The Life of the Philosopher Isidoros", written about a hundred years after Hypatia's death, Damascius gives another reason:

> So it happened one day that Cyril, the bishop of the opposition sect [of Christianity], passed by Hypatia's house and saw a large crowd of people and horses at her door. Some arrived, others left, and still others stood around. When he asked the reason for the crowd and the reason for the commotion, her followers told him that it was the house of the philosopher Hypatia and that she was about to greet them. When Cyril learned this, he was so filled with envy that he immediately began to plan her murder, and the most heinous form of murder.

The first step towards Hypatia's elimination was defamation. According to the contemporary witness Socrates Scholasticus, in a reversal of the facts, it was spread about that Hypatia, as an advisor to Orestes, was thwarting the reconciliation between spiritual and secular power in Alexandria. Two centuries later, one of Cyril's successors in the Egyptian episcopate, John of Nikiu, no longer doubted the truth of this rumour. This was the seventh century, when Christianity had long since triumphed across the board. Shortly before giving a detailed account of how Jews had crucified a small child, John writes:

> And at that time there appeared in Alexandria a philosopher, a pagan named Hypatia, who at all times had devoted herself to magic, astrolabes and musical instruments, and beguiled many people by her satanic wiles. And the governor of the city honoured her exceedingly; for she had beguiled him by her magic. And he ceased to attend church, as had been his custom (...) And he not

only did this, but drew many believers to her, and he himself received the unbelievers into his house.[7]

Under the leadership of a certain Petros, who held the rank of lector in the church, an angry mob of Christian fanatics ambushed Hypatia. The mob seized the 60-year-old philosopher, took her to the church of Kaisarion, stripped her naked and killed her with *ostraka*, or shards or roof tiles. Then they tore the body to pieces and burned it. The brutal act remained without legal consequences, for the case against the murderers was dismissed. In the contemporary records of Damascius it reads like this:

For when Hypatia stepped out of her house, as was her wont, a band of ruthless and cruel men, who feared neither divine punishment nor human vengeance, attacked her and struck her down, committing a monstrous and shameful act against her fatherland. The emperor was angry, and he would probably have avenged them if Aedesius [presumably the investigating official] *had not been bribed. So the emperor remitted the punishment from the murderers and drew it on himself and his family. His grandson paid the price. The memory of these events is still vivid among the Alexandrians.*[8]

The unpunished death of Hypatia meant a severe defeat for the governor Orestes and all those who still resisted the Christian militia. The way was now clear for Cyril to go all out. On his orders, all educational institutions were looted and the books burned. Even the great library of Alexandria—what was left of it at the time—was finally destroyed. There followed an exodus of Hellenistic intellectuals and artists from the city that had been the centre of the world of learning for seven hundred years. Those who remained professed the teachings of the Christian church or remained silent.

After the destruction of the Library of Alexandria, only one significant repository of ancient knowledge remained: the Palace Library of Constantinople (called Byzantium until 337 AD) with about 120 000 scrolls. A fire destroyed the last of the copies of ancient writings stored in the Christian cultural sphere in 475 AD. Now there were no longer any large collections of books in Europe. The time of the great repositories and mediators of knowledge was over.

* * *

[7] Robert Henry Charles, *The Chronicle of John, Bishop of Nikiu*. London (1916), Chaps. 84, 87.
[8] Ibid., Chap. 88f.

Hypatia's murder sent shock waves throughout the Roman Empire. Since time immemorial, philosophers had been considered practically untouchable, so the murder of a philosopher by a mob was seen as "deeply dangerous and destabilising".[9] The fact that the Christian Church officially condoned Hypatia's death was also met with incomprehension. In numerous letters to his religious brother and later "church father" Cyril the Younger, the contemporary witness Synesius describes the almost eponymous Bishop Cyril as inexperienced and misguided. Socrates Scholasticus, also a contemporary of Hypatias, wrote in his Church History:

> This woman was then the victim of certain machinations (...) The act brought great shame to Cyril and also to the Church of Alexandria.[10]

While the Christian Orestes still moved in both worlds, the murder of Hypatia deepened the rift between Christians and non-Christians. The chronicle of John of Nikiu explicitly approves of Hypatia's murder; his predecessor in the Alexandrian bishop's seat and instigator of the murder, Cyril, was declared a saint. On the other hand, some future Neoplatonists, including Damascius, opposed Christianity ever more passionately.

Nevertheless, some late-antique thinkers made the attempt, in good philosophical tradition, to unite the different world views—including Christianity—into a coherent synthesis. But this attempt also failed. In 529 AD, the philosophical schools of Greece were closed by law, including the Academy in Athens founded by Plato. In the same year, the first monastic order was founded, the Benedictine Order. The monasteries would now hold a monopoly on intellectual education in Europe until well into the Middle Ages. Only a few texts circulated; what was permitted and how it was to be interpreted was the sole responsibility of the Church. This was followed by a ban on teaching non-Christians and the persecution of pagan "grammarians, rhetors, doctors and jurists". Again and again the pyres blazed, burning non-Christian books and sometimes scholars too. Due to the lack of written evidence, it became impossible for Christian believers to study the ancient authors. This brought a definitive end to free thought and unbiased engagement with science. Almost a thousand years of dogmatism were to follow.

* * *

---

[9] Edward J. Watts, *Hypatia—The Life and Legend of an Ancient Philosopher*. Oxford University Press (2017).

[10] Quoted in: Arnulf Zitelmann, *Hypatia*. Basel: Beltz & Gelberg Weinheim (1988), pp. 269f.

Hypatia was an exceptional figure, not only as a scientist of the highest order, but also as a woman whose achievements had earned her great prestige in a world dominated by men. Today's historians see her death as a turning point that marked the transition from the ancient to the Christian age—and thus the beginning of a millennium of efforts to restrict education and oppose scientific thinking.

The memory of Hypatia was almost completely lost. But her mathematics was copied and passed on over the centuries, unrecognised and anonymous. In the end, the person behind this mathematics became visible again through research. What we know of Hypatia today is incomplete, but it is enough to recognise in her one of the most influential mathematicians in world history, and one who deserves to be more famous.

2

Hildegard of Bingen (ca. 1098–1179): Building Bridges Between Mysticism and Science

From the decline of Alexandria's knowledge culture until the early twelfth century, a dogmatic, anti-scientific spirit prevailed in Europe, leaving no room for critical discussion and the testing of doctrines through trial and error. Knowledge served the sole purpose of underpinning the sacred teachings of the Catholic Church. The fact that the statements of the church authorities were sometimes contradictory was studiously overlooked. Things were different in the Arab culture. It remained Aristotelian and reached its scientific heyday between about 800 and 1250.

With Peter Abelard (1079–1142), Christian theology returned to rational thinking again for the first time. The French theologian defined knowledge in a way quite contrary to the spirit of the time, as something that should have a meaning and purpose of its own. It is precisely in this "stubbornness" of thinking that lies the key to modern science: our knowledge of the world should not serve any external authority, but should be an end itself, meeting our desire to attain real knowledge. Abelard was the first to dare to point out the obvious contradictions in the Christian explanation of the world and to confront them rationally. He said that it was normal to make mistakes in thinking and that even supposedly eternally valid truths should always be open to re-examination. Abelard was thus the first to formulate the ideal of modern scientific thinking: truth is no longer just obvious, but must be sought.

L. Jaeger, *Women of Genius in Science*, https://doi.org/10.1007/978-3-031-23926-7_2

Adopting this attitude, he made accessible to theologians—that is, to philosophers anchored in Christianity—a facet of thought that had been buried for so many centuries: doubt. He writes:

> For it is through doubting that we arrive at enquiry; it is in enquiry that we grasp the truth.[1]

When Abelard began to speak scientifically about God, this implied a new emphasis on human individuality in both thought and action. For some historians, this was what triggered the "humanist renaissance of the twelfth century".

In this early phase of the new European science, another woman played a decisive role alongside Peter Abelard: Hildegard of Bingen. While her contemporary Abelard was provocative and eventually got into a fatal conflict with the highly respected and influential Abbot Bernard of Clairvaux, Hildegard of Bingen sought consensus with the Church. Nothing was further from her mind than rebelling against church leaders. And yet she was stubborn in her own unique way.

* * *

Hildegard was born in 1098 near Bad Kreuznach. As the tenth child of her noble parents, her life path was preordained: according to the "tenth child" tradition, she was to dedicate her life to the Church. At the age of eight, she was entrusted to Jutta von Sponheim, who was twelve years older. In 1112, Jutta, Hildegard, then about 14 years old, and another noble girl moved into a hermitage on the nearby Disibodenberg, where Benedictine monks had founded a monastery four years earlier. The word "hermitage" is probably to be taken literally: recluses and hermits were actually walled in in those days. Hildegard spent the next few years with Jutta and the second child locked in a room on the Disibodenberg. Their time was filled with praying—eight times a day and even at night—but also Latin lessons and needlework. Through a small window, the three could take part in services, while through another they were served with food. These openings in the wall were their only contact with the outside world.

At the age of about 15, Hildegard took her Benedictine vows. The ascetics enjoyed a high reputation, and more and more visitors would come to the hermitage to seek spiritual advice. Some women stayed and joined the group. However, the emergence of a separate women's convent did not go

---

[1] Peter Abelard, *Sic et non (Yes and No)*, new edition. Frankfurt: Minerva (1981).

smoothly—on the one hand, the convent benefitted from the gifts of visitors, but on the other hand, the spiritual success can be assumed to have given the women a growing self-confidence. From the three women, whose lives had literally depended on the donations of the monks, the hermitage became a convent of about twenty women who were ready to make certain demands. In 1136, Jutta von Sponheim died; by then Hildegard had spent almost a quarter of a century with her teacher in the hermitage. She thus took over the leadership of the convent and relaxed some of the ascetic rules established by Jutta, which were far too strict for her liking. This once again intensified the confrontation with the Benedictines on the Disibodenberg.

A few years later Hildegard prepared to move the convent from the Disibodenberg to the Rupertsberg near Bingen, where a new monastery was being built. The Benedictines wanted to prevent this separation, because if Hildegard and the other women were to leave the monastery complex, it would become overly quiet again on the Disibodenberg—and they would have to part with some of the accumulated treasures. However, Hildegard prevailed and moved to Bingen in 1151. The woman who had up to then been locked away for a large part of her life undertook four missionary journeys, from Cologne to the Swabian Alb, and from Lorraine to Würzburg.

* * *

Hildegard of Bingen's long and eventful life was accompanied by hallucinatory phenomena. Even in childhood, she had suffered strong visions, which were understood during her lifetime to be due to the intervention of higher powers. The British neurologist Oliver Sacks, however, suspects that they were caused by a scotoma. This disease is accompanied by a partial restriction of the visual field, often combined with the perception of flashes of light, altered colours, and wandering black spots.

In 1141, when Hildegard was still prioress at the women's convent on Disibodenberg, her visions had become more frequent. She suspected that they were a mandate from God to record her experiences and pass them on. To be sure of drawing the right conclusion, she turned to Bernard of Clairvaux. In the same year, the latter had ensured that Peter Abelard would be put out of action and his teachings banned; a few years later he would call for the first crusade in history. Bernard of Clairvaux replied:

We rejoice with you in the grace of God that is in you. And as for us, we exhort and beseech thee to regard it as grace and to respond to it with all the loving power of humility and devotion.[2]

A panel of theologians also confirmed the authenticity of Hildegard's visions. This extremely positive response from such powerful representatives of the Church once again reinforced Hildegard's fame. A monk was commissioned to help her write down her visions. In her first work *Scivias* ("Know the Ways"), Hildegard describes 26 prophetic visions. The original manuscript, with 35 miniatures drawn by herself, was lost during the Second World War.

At the synod in Trier in 1147, Pope Eugene III officially recognised Hildegard's visions and allowed *Scivias* to be reproduced. Her book found its way into many libraries, and she was soon exchanging letters with the most powerful men of her time; of this correspondence, some 300 documents have survived. There is a letter from Emperor Barbarossa to Hildegard which suggests that she met with him personally as an advisor, but its authenticity is disputed. Hildegard was also held in high esteem by ordinary people. Her empathetic appeals, in which she refers to her visions, were extremely popular.

Altogether Hildegard published three works in which she explained her often enigmatic visions.

- *Scivias* was written between 1141 and 1151. In this book Hildegard takes her visions as the basis for a theological overview that conforms to Church teaching in all essential points.
- The *Liber Vitae Meritorum* ("Book of the Merits of Life") is a visionary ethic in which Hildegard contrasts 35 vices and virtues. It was written between 1158 and 1163.
- She wrote the *Liber Divinorum Operum* ("Book of Divine Works", also known as *De operatione Dei*, between 1163/64 and 1173/74. When it was completed, she was already over seventy-five years old. In ten visions, she conveys the sum of all her experiences and concludes that body and soul, world and church, nature and grace are the responsibility of man.

* * *

Already in these theological disputes Hildegard demonstrates an unusual independence of thought. In her other writings, in which she deals in detail

---

[2] Hildegard von Bingen. Translated and edited by Walburga Storch, *Im Feuer der Taube: die Briefe.* Augsburg: Pattloch (1997).

with medicine, physiology, botany, and natural history, she goes one step further. In her time, there was no separate category called "science"; the accepted theories of the human body, the movement of the heavenly bodies, and man's place in the universe were combinations of theological and natural philosophical ideas, enriched with magical or mystical thinking. Hildegard of Bingen was also undoubtedly a mystic, but like Peter Abelard, she ventured into an area that had been orphaned in Europe for centuries: rationality. That is why, in addition to prophecies and allegorical treatises, her work contains genuine observations of plants and animals, but also of people and their diseases. In these passages, real science shines through for the first time since late antiquity. While her learned contemporaries did nothing but restate the statements decreed to be true by the Church, rolling them out again and again, Hildegard relied on her own perception regarding these points.

> I examine everything that happens in the greatest detail. I am my own witness to it when I sum up all things to the best of my ability according to my guidelines. What I get to see and know in this way, why should I not let that be a credit to myself?[3]

It is true that a supposed higher power also plays a role in Hildegard's scientific and medical works, for in her view everything on Earth was created by God to serve man, and if God refuses to help, even the best remedy will not be able to save the patient. But her knowledge springs from her own first-hand experience. Running the herb garden and the monastery's infirmary, she combined the spirituality of "spiritual healing" with the physical treatment of physical illnesses. This interplay shaped Hildegard's holistic method.

She coined the word *viriditas* for the vital connection between the health of the natural world and the holistic health of the human person. With this "green power" term, she was referring to a mysterious "green" life force in plants which is vital for humans, and forms the basis of all healing.

* * *

In the period between her religious visionary books *Scivias* (until 1151) and *Liber Vitae Meritorum* (from 1158), Hildegard von Bingen summarised her knowledge of nature and medicine in two works: the *Physica* and the *Causae et Curae*. Unlike her other writings, these two books were lost early on. Historians today assume that they were originally conceived as a single

---

[3] André Rademacher (ed.), *Hildegard von Bingen, Der Mensch in der Verantwortung - Das Buch der Lebensverdienste (Liber Vitae Meritorum)*. Salzburg: Otto Müller Verlag (1986).

work of natural history, entitled *Liber subtilitatum diversarum naturarum crea-turarum* ("The Book of the Secrets of the Various Natures of Creatures"). Shortly after Hildegard's death, it was divided into two works and certain passages can be found in both works. The fact that the texts known today date from over a hundred years after her death was one of the reasons why Hilde-gard's authorship was long questioned. But since the nineteenth century, there has no longer been any doubt that both books were written by her.

She collected known treatments from various sources and re-evaluated and rearranged them; she did not develop her own medical procedures. Since healing was primarily in the hands of women at that time, but they did not leave any writings due to a lack of classical training, Hildegard of Bingen's work gives us a historically unique view of practical medieval healing. Indeed, in both books, she describes specific medical applications that were widely used in everyday life: how to stop bleeding, which herbs can help to curb vomiting and diarrhoea, or what helps with burns, broken bones, dislocations, and cuts. She also discusses in detail possible applications to animals.

The first work, the *Physica* (also called *Liber simplicis medicinae*, "Book of Simple Medicine")[4] contains nine volumes—originally there were prob-ably nine individual scrolls—describing the medicinal properties of plants, animals, minerals, and other substances.

- In the first and third volumes, Hildegard presents a total of about 300 herbs and trees, i.e., the majority of the flora known at that time. She describes their medicinal uses, but also presents plants to which she attributes no effect on humans or animals. They all have their place in God's universe.
- The second volume "On the Elements" largely agrees with the theory of the elements in her second work *Causae et Curae*.
- The fourth volume, "On the Stones", deals with the precious stones, which, in addition to their metaphorical significance for Hildegard, also possess medicinal healing power.
- In the fifth volume, "On the Fishes", she begins with a description of the animal world. Birds follow in the sixth volume, quadrupeds in the seventh, and creeping animals (snakes) in the eighth.
- The last volume, "On the Metals", concludes the work with a description of eight metals.

---

[4] Hildegard von Bingen, *Physica*, translated into English by Priscilla Throop, Healing Arts Press Rochester (1998).

The knowledge that Hildegard compiled in her *Physica* alone proves Hildegard's impressive education. It is a mixture of:

(...) her own observations, (...) symbols relating to the history of salvation, traditional household remedies, customary Germanic rites, prayer-like incantations, and incantations and familiar recipes of monastic medicine.[5]

The second work, *Causae et Curae* ("Causes and Cures"), also known as *Liber compositae medicinae* ("Book of Medicinal Compounds"), explores in nearly three hundred chapters the causes of disease, the corresponding remedies, and human physiology and psychology.[6] Since in Hildegard's time many of the ancient medical writings were not available in Latin due to lack of interest on the part of the Church and had been forgotten in Central Europe—including records by Hippocrates, Dioscorides, and Galen—Hildegard's book was one of very few available to educated healers.

- In the first two sections Hildegard gives specific instructions for bloodletting. In addition to the appropriate amount of blood to be drawn, the healer should take into account other factors, including the sex of the patient, the phase of the moon (preferably, bloodletting should be done when the moon is waning) and the location of the disease (veins near the diseased organ or body part should be used). Hildegard also describes bloodletting in animals.
- The second section of the book describes in detail the physiology and pathology of the human being. Here we find descriptions of the human organs, and the different characteristics of men and women and their diseases. The theory of temperaments, based on the division of human beings into four basic characters, is also recounted.
- In the third and fourth parts of the work, she describes the treatment of complaints and diseases in humans and animals by restoring the balance between the bodily humours.
- The fifth section deals with diagnosis and prognosis and includes instructions for checking a patient's blood, pulse, urine, and bowel movements.
- The sixth section documents a lunar horoscope to provide an additional means of predicting both disease and other medical conditions, such as conception and the outcome of pregnancy.

---

[5] Änne Bäumer-Schleinkofer, *Wisse die Wege. The Life and Work of Hildegard of Bingen. A Monograph on the Occasion of Her 900th Birthday*, Peter Lang GmbH, Internationaler Verlag der Wissenschaften (1998).

[6] Abbey of St. Hildegard (ed.), Hildegard von Bingen, *Ursprung und Behandlung von Krankheiten - Causae et Curae*, newly translated and introduced by Riha Ortrun, Beuroner Kunstverlag (2011).

In this sixth part, Hildegard emphasises, for example, the importance of boiling drinking water to prevent infections. She also gives reasons for other specific instructions, giving her explanations a scientific touch.

If a person suffers from the white spot on the eyes, take fresh ox bile while the white spot is still fresh, place it as fresh as it is directly on his eyes at night, fasten it with a bandage so that it does not slip down, and proceed in this way for three days, because the bitterness of this bile breaks up and removes the pain.[7]

The term "white spots" probably refers to styes, i.e., bacterial infections of the edge of the eyelid. An eye ointment commonly used in the Middle Ages, which contained onions and garlic in addition to ox gall, was tested in modern times with surprising success: it even kept multiply resistant strains of bacteria at bay. Two pages further on, Hildegard describes the procedure for toothache:

Of toothache: He who suffers from putrid blood or from the excretion of the brain from the teeth, takes equal quantities by weight of wormwood and vervain, boils them in a new vessel with pure, good wine, strains the boiled wine through a cloth and drinks it after adding a little sugar. When he goes to bed, he puts the still warm and boiled herbs on his jaw, where his teeth hurt, and fixes them with a cloth (...) The wine tempered with the herbs mentioned and then drunk cleanses the small vessels which extend from the cerebral membrane to the gums.[8]

Even if the anatomy mentioned does not quite correspond to the facts, both herbs have a strong anti-inflammatory effect. Hildegard assumed that God had given her the procedures in a vision, but from today's perspective she probably acquired her knowledge through extensive studies, and above all, directly through her practical work.

Hildegard of Bingen's healing science, in which the idea of unity and wholeness is of central importance, is still widely known even a thousand years later. In contrast, the fact that she was probably the first person to deal with details of female sexuality is almost unknown. While she agreed with the Church's role model that women should be subordinate to men, she also paid unprecedented attention to women. Thus, in the second part of *Causae et Curae,* she describes the female orgasm with scientific meticulousness:

---

[7] Hildegard von Bingen, *Causes and Treatment of Diseases.* Königswinter: Lempertz Klassiker (2013).
[8] Ibid.

When a woman makes love to a man, a sensation of heat in her brain, which brings sensual pleasure, imparts the taste of that pleasure during the act and calls forth the emission of the man's semen. And when the semen has fallen into place, this fierce heat emanating from her brain draws the semen to itself and holds it, and soon the woman's sexual organs contract, and all the parts that are ready to open during the time of menstruation now close, just as a strong man can hold something in his fist.[9]

It should be no surprise that her intensive study of female sexuality in the Middle Ages was largely ignored. However, her descriptions were slow to receive any attention in the modern era as well. Only since the end of the twentieth century have her discussions on this subject been taken into account through renewed interest in her as a person.

*   *   *

The basic theory in Hildegard's science is an independent mixture of Greek and Latin traditions and Christian thinking in folk medicine. The basis is a fusion between different patterns focused on the number four:

- The four-fluid theory assumes that blood, phlegm, and black and yellow bile must be kept in balance in humans. Important representatives of this view were the Greek Hippocrates and the Roman physician Galen, who distinguished between four temperaments on the basis of the respective predominant humours: melancholics determined by black bile, sanguine people characterised by red blood, phlegmatics with an excess of white mucus, and the cholerics, who flare up due to their yellow bile. Whereas the ancient physicians placed all four humours on an equal footing, Hildegard divided the quartet into the higher humours of light bile and blood and the lower humours of dark bile and phlegm.
- The theory of the four elements—fire, air, water, and earth—also originated in antiquity. Hildegard developed this view further by distinguishing between the higher, intangible, heavenly elements of fire and air and the lower, tangible, earthly elements of water and earth.
- In her doctrine of the four continents, Hildegard follows the ancient philosopher Crates of Mallus, who in the second century B.C. asserted that the Earth was spherical and described the north, which was too cold for humans, the hot south, which was also uninhabitable for humans, and the inhabited countries of Asia, Africa, and Europe as the third continent.

---

[9] Hildegard von Bingen, *Causae et Curae*, translated (English) by Manfred Pawlik and Patrick Madigan, edited by Mary Palmquist and John Kulas, Liturgical Press, Inc (1994), second chapter.

As a fourth continent, he considered a possible land mass accessible via the Atlantic Ocean, which he called "perioikoumene" ("around the Earth").

Other patterns associated with the number four considered important in the Middle Ages were morning—noon—evening—night, childhood—youth—adulthood—old age, and spring—summer—autumn—winter. Everything seemed to interlock in a mysterious way.

With the help of these quartets, Hildegard of Bingen worked out her very own, almost scientific explanation of how the human cosmos and the universe were connected. Based on her visions, she united the four humours and the four elements into the pairs light bile/fire, phlegm/water, dark bile/earth, and blood/air, which are the decisive factors for health and illness in the human body.

- Hippocrates had already taught that out-of-balance humours lead to diseases; he called this imbalance *dyscrasia*. For Hildegard, diseases arise from the inadmissible dominance of *subordinate* humours.
- If the elements and juices are balanced (*eucrasia*), the person is healthy.

The most important means for countering this problem was bloodletting. The literally medieval concept of "juices that must be in balance" was continued by medical practitioners into the early nineteenth century. US President George Washington died in 1799 from excessive bloodletting, in which almost two litres of blood were taken from him over a short period of time.

While bloodletting plays only a very minor role in today's medicine, another of Hildegard of Bingen's views seems very modern: for her, the goal of medicine was not only to cure illness, but also to enable a healthy lifestyle. Prevention rested on five pillars:

- Food and medicines: For Hildegard, there was no separation; both were means supporting life and healing. They are found in the Creation of the World among plants, fish, animals, and precious stones.
- Rest and activity: Movement and rest (in the form of prayer and meditation) should be part of a healthy balance. The well-known rule of the Benedictine monks *ora et labora*—pray and work—finds a deeper meaning here.
- Waking hours and sleep: The health-promoting regeneration of the organism requires the observance of certain rules and a certain rhythm with regard to waking hours and sleep.

- Extractions from the body: Expulsions such as bloodletting rid the body of toxins caused by overeating, dietary errors, stress, worry, fear, and disappointment by removing "bad blood".
- Strengthening mental defences: Constant self-reflection ensures health and well-being on all levels. This focusing on oneself includes the task of becoming aware of one's own strengths and weaknesses, as well as fathoming the connections between feelings, thoughts, and physical sensations.

* * *

On the one hand, Hildegard of Bingen was firmly rooted in the three Benedictine virtues of humility, obedience, and right measure; on the other hand, she was also a strong personality who developed completely new ideas and possessed the necessary authority to present her arguments in the face of controversy when necessary. In medieval society, there was usually no place for self-confident women like her, but Hildegard managed to secure the veneration of the dignitaries within the church—i.e., in an industry that was exclusively male at the time.

Towards the end of her life, however, there was still a profound and painful confrontation with her church. An excommunicated nobleman had been cared for in Hildegard's convent; when he died, he was buried in the convent cemetery. However, the Mainz cathedral chapter did not recognise that he had received the last rites after a confession and demanded that his body be buried in unconsecrated ground. Because Hildegard refused to comply with this request, it was forbidden for the monastery to celebrate services. The church doors had to remain closed, the bells could not be rung, and singing was also forbidden. Only after a few months did the Archbishop of Mainz lift the interdict. A short time later, Hildegard died at the age of 81.

Even after her death, Hildegard was revered. Just as in her lifetime, when she was both the Church's most faithful servant and a stubborn forward thinker who went beyond its boundaries, she is today both saint and non-saint at the same time. In 1228, proceedings were opened for her canonisation, but they have not been completed to this day. Almost 800 years later, in May 2012, she was nevertheless included in the calendar of saints. In October of the same year, Pope Benedict XVI even elevated her to the rank of Doctor of the Church. This honorary title puts her on a par with Augustine and Thomas Aquinas.

Hildegard of Bingen did not concern herself with mathematics at any time during her life. Her explanations are interwoven with a deep religious faith. Moreover, she was an early mystic whose methods were the exact opposite of modern science. She is nevertheless included in this book, and for a special reason: like Peter Abelard, she conveyed the very first thoughts committed to rationality. She opened the door to a dynamic process from which modern scientific methodology would eventually develop, very slowly at first, and then, from the seventeenth century onwards, ever more rapidly. Along with Peter Abelard, she is considered by many historians to be the founder of the scientific history of Europe.

3

Laura Bassi (1711–1778): The World's First Female University Professor

After Hypatia's death, it would be seven centuries before another woman, Hildegard of Bingen, asserted herself in the male-dominated domain of scholarship. After her death, another six hundred years would pass before any other important female academics would make an appearance in the scientific world. Even though the universities were growing more prestigious and knowledge was accumulating in them during this period, women were strictly forbidden in academia. Only a very few could make a name for themselves in scientific circles or win a place in the lecture halls.

- Bettisia Gozzadini's field of expertise was jurisprudence, so she was not a scientist, but we can find a place for here on this list because in 1237 she graduated from Bologna, the oldest university in the world, and was probably the first woman in history to do so. Until she was finally allowed to teach at the university from 1239, she taught privately at her home.
- Dorotea Bucca, a professor of medicine, is said to have taught in Bologna at the end of the fourteenth century, but the sources are disputed.
- At the beginning of the Enlightenment, in the first half of the seventeenth century, Anna Maria van Schurman was allowed to listen to lectures at Utrecht University, with official permission. However, she had to sit behind a curtain so that male students could not see her.
- At this time, Maria Kunitz, Elisabetha Hevelius and, somewhat later, Maria Winkelmann made a name for themselves as astronomers in Germany.

© The Author(s), under exclusive license to Springer Nature Switzerland AG 2023
L. Jaeger, *Women of Genius in Science*,
https://doi.org/10.1007/978-3-031-23926-7_3

- One of the most famous women of the seventeenth century was the British natural philosopher Margaret Cavendish. Among other things, she did not like Newton's purely mechanistic explanation of the world, but the biblical form of explanation did not appeal to her either. Instead, she developed her own explanation for the phenomena of nature, proposing a kind of "animated" matter that could arrange and rearrange itself on its own. Despite the sometimes harsh criticism from both Newtonians and clerics, she was the first woman to be invited to meetings of the *Royal Academy*, which was founded in 1660.
- In 1678, Elena Lucrezia Cornaro Piscopia was the first woman in the world to obtain a doctorate. It was actually to be a doctorate on a theological subject, but church dignitaries objected. Without further ado, since Piscopia was as well versed in philosophy as she was in theology, she did her doctorate on Aristotelian logic. Unfortunately, she did not live long enough to leave her mark on the academic world; six years after her doctorate, she died at the age of only 38.
- Sybilla Merian systematically explored the insect world without any scientific training, and undertook a research trip to the Dutch colony of Surinam from 1699 to 1701.

The strict understanding of male and female roles in European societies prevented the aforementioned personalities from meeting the intellectual giants of their time on an equal footing, so there was no way they could exert any greater influence in their fields. It was not until the beginning of the eighteenth century that a little flexibility was allowed into these entrenched conventions. During the Enlightenment, the scientific revolution rolled over Europe like a mighty wave. Natural philosophy was no longer a matter for individual scientists; the number of private scholars and university professors suddenly increased. The educated middle classes also took a greater interest in discoveries made in physics and mathematics, and issues of philosophy and natural history were passionately discussed in salons and cafés. However, only half of Europe's population could yet be actively involved in these developments.

Although studying natural sciences was out of the question for women, two of them succeeded in developing their talents and becoming important representatives of their fields: the Frenchwoman Émilie du Châtelet, to whom the following chapter is devoted, and the Italian Laura Bassi. While Émilie du Châtelet waited in vain throughout her life to be admitted to the *Académie des Sciences* in Paris, Bassi was more fortunate: she lived in the catchment area of the comparatively progressive University of Bologna, which enabled her to become the world's first natural scientist to earn a doctorate and soon after to train her own students as a professor. In addition to her sound theoretical and mathematical understanding of physics, she possessed an outstanding talent for experimental physics research. She was one of the first people to systematically address the phenomena of electricity and magnetism.

* * *

Laura Bassi was born in Bologna in 1711, the daughter of the wealthy lawyer Giuseppe Bassi and his wife Rosa Marie Cesarei. When she was five years old, her cousin Father Lorenzo Stegani began teaching her Latin, French, and mathematics. From the age of thirteen, Gaetano Tacconi, the family doctor and distinguished professor of medicine at the University of Bologna, was her tutor in philosophy, metaphysics, logic, and natural philosophy. After seven years, the two fell out, because Tacconi insisted on concentrating on Descartes' physics. But Bassi, now twenty, was more interested in Newton's science. This dispute between teacher and pupil directly reflected the conflict between two camps that divided the scientific world at the time.

- Descartes had developed an elaborate and complicated theory of cosmic phenomena: vortices of matter created by God were supposed to explain all celestial phenomena, including the formation of the Sun and planets, the planetary orbits, and also gravity. With the vortices as a medium, two distant objects, for example the Earth and the Sun, could exchange forces directly.
- Newton's universal theory of gravity, on the other hand, no longer required direct contact between the celestial bodies—or between the apple and the ground. With his formula, gravity could even be correctly calculated in mathematical form.

Which was the "right" view was hotly debated by scholars and also by educated citizens, and the arguments often went personal. The dispute was roughly comparable to the debate which raged during the Cold War in the twentieth century over the question of whether capitalism or communism would bring the greatest happiness.

In 1731, at the time of the break-up with Tacconi, Bassi was already working on her first work of natural philosophy. In 1732, she presented 49 theses on logic, metaphysics, and physics, as well as on the nature of man, in her *Philosophica Studia*. Bassi and her high level of education and intellect had previously come to the attention of the influential Archbishop of Bologna,

Prospero Lorenzini Lambertini. As her patron, he arranged a two-hour debate that took place on 17 April 1732 between Bassi and four professors. This was also her doctoral examination. This public event in Bologna's town hall became a great spectacle, as a sensation-hungry crowd turned out to see the learned woman. Bassi confidently defended her 49 theses and impressed the professors sitting opposite her with her knowledge, some of her answers causing the auditorium to cheer loudly.

After this success, everything happened very quickly: on 12 May, the University of Bologna awarded her a doctorate in philosophy (which at that time also included the natural sciences). In addition, the 21-year-old was elected to the Bologna Academy of Sciences. Never before had a woman been granted this honour, neither in Bologna nor at any other university. The enthusiasm for her achievements found expression in public celebrations and the publication of volumes of poetry in her honour. Bassi seized the opportunity and submitted a request to the senate of the University of Bologna for a teaching position, i.e., a professorship. Only six weeks after she was awarded her doctorate, she defended twelve further theses in a kind of habilitation lecture in the *Palazzo dell'Archiginnasio,* the central building of the university. These theses covered a broad range of topics from chemistry to physics, hydraulics, mathematics, mechanics, and technology. On 29 October 1732, the Senate and the University of Bologna approved Bassi's candidacy and appointed her Professor of Philosophy in December. Thus, within a single year, Laura Bassi made a career from private student to the first permanent and also paid university lecturer in history. Far beyond Bologna, she was revered as "Minerva," and thus as a symbol of the sciences.

All these extraordinary successes had taken place in an atmosphere of general exuberance and enthusiasm that had everyone carried away, including the established bodies. Now it was time for the doubters to have their say. As progressive and important as the University of Bologna was in comparison to other European universities, there too the prevailing view was that women were unwelcome in the university environment. Unlike her male colleagues of the same rank, Bassi had to have each of her lectures explicitly approved by the magistrate. This was no mere formality, for permission was all too often denied, to her great displeasure. While her rights were curtailed, she was expected to fulfil in full all the duties of a professor. Among other things, the Senate expected her to attend various public events in order to benefit from her enormous reputation. In part, Bassi met this demand quite gladly. For example, from 1734, she attended with great interest the annual "Carnival Anatomies," during which corpses were publicly dissected in the *Teatro anatomico* of the University of Bologna.

$*\ \ *\ \ *$

Laura Bassi was the only surviving child of her parents and, after her father's death, had sufficient means to make independent decisions. When it came to marriage, she would only consider someone who would not restrict her freedoms, and certainly not someone who would force her to give up her academic career. To the astonishment and displeasure of the people of Bologna, who were extremely proud of their scholarly figurehead of, she married the not very wealthy Giuseppe Veratti, a doctor of medicine and lecturer in anatomy at the University of Bologna, in 1738. She wrote about her expectations in a letter that year:

> (…) and that is why I chose a person who, like me, was progressing along the path of science, and who I could be sure, through many years of experience, would not try to dissuade me.

Over the next fourteen years, the couple had eight children, five of whom reached adulthood. This did not stop Laura Bassi—who was now actually called Veratti, but continued to be known as Bassi—from lecturing. Her husband, being a doctor, was more practically inclined, and he inspired her to become more involved in experimental physics. Together they set up a private laboratory and conducted numerous experiments there.

Outside her private life, however, her efforts to gain acceptance and equality proved to be a constant battle against windmills. The ranks of her powerful friends and patrons did nothing to change this. In 1739, her request to be granted full rights to her teaching post was again denied, despite the explicit support of prominent patrons such as Archbishop Lambertini and Flamino Scarselli, the secretary of the Bolognese ambassador to the papal court. She scored a partial success in 1740 when at last she was allowed to give private lessons at her home. This permission gave her the opportunity to focus on her own choice of subjects outside the university curriculum.

Another success, although it was not achieved without setbacks along the way, was her acceptance into the group known as the *Benedettini*. In 1745, her patron Lambertini, by then Pope Benedict XIV, reorganised the Academy of Sciences of Bologna and initiated the foundation of a select group of twenty-five scholars. Bologna was one of the best and most famous universities, so the *Benedettini* were among the most important scholars in Europe. At least once a year they were to present their research results and thus provide a cross-section of current topics in the sciences. Laura Bassi did her utmost to be included in this select flock, but met with categorical resistance from most of her male colleagues who had already been chosen. Benedict XIV, one of the most powerful men in Christendom, could do nothing against the scholars insisting on their privileges and had to offer a compromise: Bassi would indeed become the twenty-fifth member of the group, but without the same voting rights as her colleagues.

* * *

Laura Bassi was particularly interested in Newtonian physics. While the in some respects old-fashioned teachings of Descartes and Galileo were still on the curriculum at the University of Bologna, she turned to the red-hot topics of gravitational force and the dynamics of fluids like the air. She was also fascinated by the mysteries of electricity and magnetism, which at that time had not yet been recognised as interrelated phenomena. She was a great theoretician, with an excellent grasp of mathematics, and in time became an first-rate practitioner, conducting numerous novel experiments in her private laboratory. Curious and courageous, she broke away from old explanatory patterns when devising theories for what she observed. Soon students from all over Italy and even other parts of Europe were coming to Bologna to hear Bassi's lectures and witness her experiments. Her courses, each lasting eight months, offered participants far more comprehensive instruction than the university's courses—neither Newton's physics nor Franklin's electricity were on the official curriculum in Bologna.

In 1755, Bassi founded a school and discussion centre for experimental physics. One of her students was the young Alessandro Volta, who later made revolutionary discoveries in the field of electricity and constructed the first electricity-generating machine. Established natural scientists also found their way into Laura Bassi's house, including Abbé Nollet, who wrote his extensive work on experimental physics a few years later and became the first professor in this field in France. Another was Lazzaro Spallanzani, who discovered the connection between the diameter of blood vessels and the speed of blood flow, and later made important experimental contributions to the study of bodily function and reproduction in animals and became a respected member of many scientific societies and academies. Spallanzani once stated that he would never have become an experimenter if it had not been for his studies with Laura Bassi.

The practical side of physics, however, has always had one major drawback: although Bassi drew one of the highest salaries at the university, experimentation is a very expensive business. Unlike many of her colleagues, she had to pay for equipment and materials out of her own pocket. On 16 July 1755, she wrote in a letter to Scarselli:

As far as my physical experiments are concerned, I am almost desperate given the fact that the constant costs I incur require some form of support if I am to push them forward and perfect them.

It was to take another twenty years or so before the university granted her any funding for her experiments.

* * *

Laura Bassi was fascinated by the way air and liquids behave under pressure and in motion. The title of her first lecture, with which she obtained her doctorate in 1732, already indicates her interest in this topic: "*De aqua corpore naturali elemento aliorum corporum parte universi*" (On water bodies as the natural element of all other bodies in the universe). Later, in the second half of the 1740s, there followed observations such as "On the compression of air," "On the bubbles observed in free-flowing liquids," and "On air bubbles escaping from liquids." Many of Bassi's experiments concerned the Boyle–Mariotte law, which describes the constancy of the product of pressure and volume for ideal gases; among other things, she researched the influence of temperature on gas properties. Her findings had some partial practical applications. For example, she developed a mathematical procedure with which hydraulic engineers could calculate the optimal size and position of pipe openings under water under ideal conditions; however, the influence of turbulence would only be mastered mathematically much later.

She was also fascinated by electricity because, as with gravity, invisible forces were at work. The phenomenon, which at that time had hardly seen any serious research, was nevertheless known to a wide audience as a spectacle: in public shows, test subjects were charged with electricity so that they could ignite gunpowder and alcohol, produce sparks, or give spectators small electric shocks. But Bassi wanted to get to the bottom of it. No one before had systematically studied the nature of electricity and magnetism. These phenomena had been known since antiquity, but scholars did not know much about them. Even Newton had only mentioned them in passing.

Decades before Luigi Galvani undertook his famous experiments with frog legs, Bassi and Veratti were devising fundamental scientific experiments on electricity. While Veratti was researching therapeutic applications against headaches, arthritis, rheumatism, watery eyes, and even herpes infections, Bassi wanted to learn more about the phenomenon itself. As early as 1752, on her initiative, the world's first lightning conductor had been installed on the roof of the Academy of Sciences in Bologna—in that year, Benjamin Franklin had published his theory that lightning was nothing more than electricity that could be discharged in a targeted manner. However, the lightning rod had to be taken down again after only a short time as the population were afraid that it would attract lightning. Bassi's experiments in the 1760s reinforced her view that electricity was a fluid, although the existence of "flowing" electrons was not proven until 150 years later.

* * *

Due to her administrative duties, family responsibilities as the mother of eight children, and frequent illnesses, mostly related to pregnancy and childbirth, Bassi published very little in writing. She never wrote a book summarising her broad knowledge, and of the total of 28 papers, mostly on physics and hydraulics, which she submitted to the Academy on the occasion of her annual lectures, very few have survived.[1] Apart from her doctoral thesis of 1732, only four of her papers were officially published:

- *De aeris compressione* ("On Air Pressure", 1745)
- *De problemate quodam hydrometrico* ("On Certain Problems of Hydrometry", 1757)
- *De problemate quodam mechanico* ("On Certain Problems of Mechanics", 1757)
- *De immixto fluidis aere* ("On the mixing of liquids from the air", published posthumously in 1792)

It is above all her correspondence with the most important European personalities that testifies to her influence on science. Her pen pals included the Geneva natural scientist, philosopher, and lawyer Charles Bonnet, the physicist Alessandro Volta, and the Italian polymath Paolo Frisi. Voltaire was also in regular contact with her. When he sought admission to the Academy of Sciences in Bologna, just like Bassi, he wrote to her:

> There is no Bassi in London, and I would be much happier to be admitted to your Academy of Bologna than to that of the English, even if it produced a Newton.[2]

---

[1] An overview of her scholarly works is provided by Domenico Piani in *Catalogo dei Lavori dell'Antica Accademia, raccolti sotto i singoli autori* (1852). In more recent versions in: A. Elena, *In lode della filosofessa di Bologna: An Introduction to Laura Bassi*, Isis, 82, 3, pp. 510–518. 10.1086/35583 (https://www.journals.uchicago.edu/doi/10.1086/355839); and also B. Ceranski, *And She Fears No One: The Physicist Laura Bassi (1711–1778)*. Campus-Verlag, Frankfurt a. M. (1996).

[2] P. Findlen, *Science as a Career in Enlightenment Italy: The Strategies of Laura Bassi*, University of California Press, Vol. 84, No. 3, pp. 441–469. https://www.journals.uchicago.edu/doi/10.1086/356547.

Perhaps this flattery contributed to his later actually becoming a member of the Academy.

In the course of her scientific career, which lasted for many years, Laura Bassi had to be careful not to be torn between the poles of the great recognition she enjoyed in the scientific community and the constant rejections within that same community. It must have been particularly painful for her that her alma mater, the highly prestigious and progressive University of Bologna, took so long to evaluate her solely according to her scientific merits. Her high salary could not hide the fact that she was not really considered as an equal member of the professorial body.

It was not until 1776, two years before her death, that her achievements were finally recognised with the conferral of the chair of natural sciences in Bologna. Even then, this leading position for experimental physics had first been offered to her husband. Fortunately, Veratti, who was a medical doctor and not a scientist, had turned down the post in favour of his wife, and after her appointment, he worked as her assistant. Her massive support for the sciences, and not least for the education of the next generation in her own home, as well as the reputation of her own scientific work, had left the decision-makers little choice. At last, she was able to teach her students at one of the most famous universities in the West, independently of the outdated ideas to which many of her colleagues and superiors still adhered.

* * *

Laura Bassi was highly educated and well versed in the canons of philosophy and physics. Beyond these scientific interests, she was familiar with both classical literature and the modern literature of France and Italy and she was a popular member of many literary circles.

In science, she did not blindly follow any particular school of thought, but made up her own mind on every subject. She remained intellectually agile until the end of her life.

- The 49 theses she presented to the University of Bologna at the age of twenty were still strongly influenced by Aristotle from a metaphysical point of view, i.e., with regard to the basic principles and causes of everything that exists. For example, in accordance with this ancient thinker, she refers to the four causes that come together in the becoming of an object: the *causa materialis*, the material, the *causa efficiens*, the agent that acts on this material, the *causa formalis*, the form that the object takes, and the *causa finalis*, the purpose for which the object is intended.

- She followed the teachings of Descartes in explaining our capacity for cognition. Essentially, she shared his view that the consciousness of our self proves the truthfulness of our sensory perceptions. However, she rejected the separation of mind and matter as taught by Descartes.
- In the field of the natural sciences and natural philosophy, she followed Newton, whose physics of motion and gravitation she found more convincing. Like him, she assumed that all phenomena in the world could be explained mechanistically—including those in biology and chemistry. But even Newton was not safe from her criticism: she did not follow his ideas on the work of God in the world, with which he explained the movement of the planets. Nor had she much time for his obscure and irrational experiments in alchemy.

Her independence of thought made Laura Bassi one of the key figures of the Enlightenment in Italy, a country where scientific thought was still strongly influenced by Galileo in her day. She paved the way for the introduction of Newton's mechanistic ideas in Italian thinking. The resulting reorientation of the natural sciences in her country was one of her most important achievements.

Laura Bassi died in 1778 at the age of 66. Her health had suffered, among other things, due to her numerous pregnancies and sometimes difficult births. The cause of death was given as an "attack in the chest," probably a heart attack.

4

Émilie Du Châtelet (1706–1749): Mastermind of the Early Enlightenment

When Newton gave the world his laws of mechanics in 1687, the scientific revolution was born. With them, the way was cleared for a completely new world view in which a divine being no longer mysteriously determined the flight of a cannonball or the load limit of a loading crane, but events could be scientifically explained and even predicted. Half a century earlier, Galileo and Kepler had already described this radically new idea, but it had only just found fertile ground.

Newton became a national hero in his native England, but it took many decades before his thoughts made the leap to mainland Europe. Even the later editions of his work in 1713 and 1726 did little to change the fact that his *Philosophiae Naturalis Principia Mathematica* (the *Principia* for short) was only understood by a few experts. It was a woman who translated the *Principia* from Latin into French and made it comprehensible to a wider readership by means of quite independent improvements to the mathematical presentation. Her name was Émilie du Châtelet. Barely sixty years after the first edition of Newton's work, it was Châtelet who enabled the breakthrough and further development of the new physics on the European mainland, thus fuelling the Enlightenment. In doing so, she also followed her own innovative scientific thoughts, which went beyond Newton and took her 100 years ahead of the field in some cases.

As a great thinker and exceptional scholar of the eighteenth century, Émilie du Châtelet exchanged ideas with almost all the great scientists of her day. Frederick the Great, under whose reign the Royal Academy of Sciences in

© The Author(s), under exclusive license to Springer Nature
Switzerland AG 2023
L. Jaeger, *Women of Genius in Science*,
https://doi.org/10.1007/978-3-031-23926-7_4

Berlin experienced its first heyday, was also an admirer. The correspondence between the Prussian king and the French scholar, which lasted from 1738 to 1744, is a remarkable testimony to du Châtelet's self-confidence, which was unusual for a woman in those days:

> Judge me by my merits or by the lack of them; but do not regard me merely as the retinue of, say, that great general or that deserving scholar, that star of the French court or that famous poet. I am a man of my own and responsible to myself alone for all that I am or do. There may be metaphysicians and philosophers whose knowledge is greater than mine; I have not yet met them. But even they are but weak, flawed men, and if I add up my gifts, I may well say that I am inferior to no one.

* * *

Émilie du Châtelet was born on 17 December 1706; her father was an officer at the French court who had bought himself a lower title of nobility in 1699. The talents of his only child were recognised early on; when she was about ten years old, teachers of mathematics, literature, languages, and science came to the house. By the age of twelve, Émilie was fluent in Latin, Italian, Greek, and German. As her parents were socially active and ran a salon in Paris according to custom, she came into contact with many celebrities of the time during her childhood, including the influential French Enlightenment philosopher Jean-Baptiste Rousseau.

At the age of 18, Émilie married into the higher nobility. Her marriage to the Marquis Florent Claude du Chastellet (the spelling of this name was later changed to Châtelet), who was twelve years her senior, produced three children over the following seven years. Now that the family succession was assured, her husband accepted, as was customary at the time, that his wife had affairs during his absence. Since the marquis, as general and diplomat to the king, was rarely at home, Émilie du Châtelet had ample opportunity to lead her own life. In 1733, at the age of 26, she took up mathematical studies.

She soon outstripped her teachers and set out to find a mathematician from whom she could still learn something. Her choice fell on Alexis Clairaut, seven years her junior, who was considered a mathematical prodigy; he became her teacher and, for a time at least, her lover.

In the long term, however, her heart belonged to another. Shortly after the birth of her third child, she had begun a passionate affair with the philosopher and man of letters François-Marie Arouet l(e) J(eune)—better known by the anagram Voltaire. Each appreciated the other's sharp mind. Only a year after their relationship began, Voltaire's latest publication led to a warrant being issued for his arrest. He had to leave Paris and found shelter in one of du Châtelet's husband's estates. Voltaire set up home in the chateau of Cirey in the Champagne region, because from here he could have fled to nearby countries if necessary. Du Châtelet gave up her previous social life and followed him to Cirey, where she increasingly devoted herself to mathematics and later to physics. The Marquis had no objections; on the contrary, he was pleased that his wife's lover, who had become very wealthy, through dubious investment transactions among other things, had the château renovated and extended at his own expense. They did not interfere with each other; all three were regularly present in Cirey at the same time. Voltaire and du Châtelet had numerous other affairs with other partners, but their intense relationship lasted fifteen years.

The château in Champagne quickly became very popular, as the famous literary figure Voltaire had his plays performed in the private theatre he built. But Cirey also developed into a centre for science. A new wing of the building housed a research facility with the latest laboratory equipment from London, and the book collection was on a par with the library of the *Académie des Sciences* in Paris.

* * *

Here was an ideal image of the enlightened age: together, du Châtelet and Voltaire discussed philosophical topics, devoted themselves to the fine arts, and conducted physical experiments. But even this relationship was apparently not entirely free of vanity. Both wished to win the regularly awarded prize of the Paris *Académie des Sciences*, hub of current research topics. In 1738, scholars were invited to submit scientific papers on the subject of the "nature of fire". Du Châtelet knew that Voltaire was writing his own contribution, which was not very original and was mainly based on the ideas of the Dutch physicist and mathematician Gravesande and his colleague Musschenbroek. She worked secretly at night in the castle laboratory so as not to give her lover any insight into her findings. But the submission of her work was also to turn into a farce: Du Châtelet, although she was already in lively exchange with the most important scientists of her time and could be sure of their admiration, was not allowed as a woman to attend the meetings of the *Académie* or to submit a paper there.

She had experienced a similar situation a few years earlier. When she had tried to enter the Café Gradot in Paris, the most important meeting place for the leading mathematicians of the day, she had immediately been shown the door. A week later, she appeared again at the café, this time disguised as a man. She could not have deceived anyone with her appearance, and that wasn't her aim. Her appearance was taken as it was meant to be, as a protest against the rule excluding women from the scientific establishment. The mathematicians were amused and let her in. When she submitted her work on fire to the *Académie des Sciences,* she again disguised herself as a man, and again with success. After overcoming this hurdle, her manuscript had the same chances as her competitors, because one of the rules of the Academy was that the competition manuscripts would be evaluated anonymously.

Neither du Châtelet nor Voltaire won the prize that year. It went to the Swiss mathematician Leonhard Euler, who incidentally won it twelve times in total, and to two other scientists. Only now did du Châtelet reveal to her friend Voltaire that she had also submitted an entry. Voltaire's reaction is not known, but that of the *Académie des Sciences* is. They were so taken with her *Dissertation sur la nature et la propagation du feu* (Dissertation on the Nature and Propagation of Fire) that it was published under her name at the expense of the *Académie*, making her contribution the first scientific work by a woman to be published officially and with the support of the responsible institution. In her work, du Châtelet came to some interesting conclusions about the nature of light, which was still controversial at the time. Among other things, she suggested the following:

- Light has an enormously high speed. If it had a mass and it hit the Earth, the consequence, would be devastating. She concluded that, since such effects are not observed, light must be massless. This is exactly what is confirmed by today's quantum theory.
- There could be more colours than those recognised by the human eye. Infrared radiation and UV light were not discovered until 170 years later.

* * *

Despite her successes and increasing network in the scientific world, du Châtelet repeatedly faced obstacles. At the end of the 1730s, she translated Mandeville's fable "Of the Bees" from English into French. The rather strange work on morality had hit European societies like a bomb with its subtitle "Private Vices as Public Advantages". Du Châtelet not only translated, she also commented and edited the text, omitting certain sections and adding material in other places. Like modern scholars, she clearly marked the changed passages in each case. In her preface, as so often, she took a clear position on the discrimination of women:

I feel the full weight of the prejudice which so generally excludes us from the sciences; it is one of the contradictions in life which has always astonished me when I see that the law allows us to determine the destiny of great nations, but that there is no place where we are trained to think (...) Let the reader reflect why at no time in the course of so many centuries has a good tragedy, a good poem, a distinguished narrative, a beautiful painting, a good book on physics been produced by a woman. Why these creatures, whose minds seem in every respect to be similar to men's, are stopped by some irresistible force, but until they are, women will have reason to protest against their education. (...) I am convinced that many women are either unaware of their talents because their education is deficient, or that they conceal them for lack of intellectual courage because of prejudice. My own experience confirms this. Chance introduced me to literary men who extended to me the hand of friendship. (...) I then began to believe that I was a being with a mind (...)

Her contentiousness was particularly evident in the fierce conflict with the influential mathematician, astronomer, and geophysicist Jean-Jacques Dortous de Mairan. The debate ignited over du Châtelet's major work on physics, the *Institutions de physique,* which she published in the spring of 1740 at the age of 34. In it, she gave an overview of the theories of Descartes, Newton, and Leibniz, and thus of the state of the art in physics and astronomy. In the last part of her book, she attacked Mairan's work of 1728, in which he specifically adhered to Descartes' vortex theory of 1644: all space should be filled with a circularly moving ether that pushes all matter into the centre of the vortices – this was Descartes' explanation for gravity. Du Châtelet did not much like this theory, which still had a major influence on physics in France. She found Newton's more modern ideas much more plausible. They had been introduced to her by the French mathematician Pierre-Louis Moreau de Maupertuis, who was also her lover for a time.

But du Châtelet was also highly critical of certain aspects of Newtonian physics, which had only partially detached itself from Descartes' ideas. At her request, the German mathematician Samuel König had introduced her to the theories of the German polymath Leibniz. She found his views far more convincing than those of Descartes or Newton—and the first chapter of her book is still one of the clearest presentations of Leibniz's teachings in French.

Du Châtelet thus knew everything she needed to compare and evaluate the views of the three heavyweights of European science, Descartes, Newton, and Leibniz, in her book *Institutions de physique.* Her book was translated into Italian and German (in German the title was: "Der Frau Marquisinn von Chastellet Naturlehre an Ihren Sohn zur Physik"), and it was so successful that her teacher Samuel König tried to impersonate its author. This betrayal was only one of the disappointments du Châtelet had to deal with. For now, Dortous de Mairan entered the stage.

It was a strong point of du Châtelet's book that, like so much of what she wrote, it led to heated debates among Europe's scholars—discussion is, after all, a key part of doing science. But the counterattacks by the staunch Cartesian de Mairan exceeded the bounds of decency. He would not listen to his colleagues who advised him to ignore du Châtelet's book so as not to bring it any more public attention. Instead, he became personal and took a shot at du Châtelet. He felt it was beneath his dignity to have to discuss such things with a woman and advised her, given her lack of knowledge, to stay at home, which was her place after all.[1]

But du Châtelet would not even think of giving in, and published the corresponding arguments of the debate in the following edition of her *Institutions*, which appeared in 1742, so that each reader could make up their own mind.[2] She received support from her Newtonian teacher, Maupertuis, among others. He publicly praised her work and took a clear position: "She is right in the matter and in the form."

In the third edition of her work, which was published in 1744, du Châtelet had part of her answers deleted again and wrote:

I am not secretary of the academy, but I am right, and that is worth all the titles.

This, of course, referred to Dortous de Mairan, who was the chief secretary of the *Académie des Sciences*. This self-confidence considerably increased her notoriety and prestige, and it was precisely what de Mairan's colleagues had feared. One of the many readers of her book was the young Immanuel Kant. In 1747, the 23-year-old first came to public attention with his work "Thoughts on the True Estimation of Living Forces", the subject of which was the dispute between du Châtelet and Dortous de Mairan.

* * *

Let us say a word about this dispute. At that time, force and energy were still used as synonyms, even Leibniz and Newton had not yet distinguished between these physical quantities, although they differ fundamentally from

---

[1] *Lettre de M. de Mairan à Mme la marquise du Chastellet sur la question des forces vives, en réponse aux objections qu'elle lui fait sur ce sujet dans ses Institutions de physique' sur Gallica* (1741).

[2] *Réponse de Mme la marquise Du Chastelet à la lettre que M. de Mairan, secrétaire perpétuel de l'Académie royale des sciences, lui a écrite, le 18 février 1741, sur la question des forces vives* [archive] (16 mars 1741).

one another. Du Châtelet was the first person in history to explain the concept of energy as something independent of force. Based on her own empirical studies, she quantified the connection between energy and force, distance, mass, and spatial position, and was thus a century ahead of her time.

A particularly contentious point in the discussion about forces and energies was the phenomenon of friction. Newton had imagined that all motion in the world, and thus also energy, must continue to decrease due to frictional forces, and that God alone could, so to speak, wind the clock up again. He also held the view that the energy of motion was proportional to the velocity (v) and the mass (m). However, according to today's knowledge, the equation $E = mv$ in his formulation describes neither force nor energy, but the momentum (p) of the system: $p = mv$. Leibniz and his followers, on the other hand, were of the opinion that "force" (meaning energy) was not lost through friction, but only transformed. Leibniz had also already made assumptions that the kinetic energy (for him still "force") was proportional to the *square* of the velocity, i.e., $E = mv^2$. He called this "force" *vis viva*, or living force. Du Châtelet supported this assumption by bringing into play a possible principle of equality of energies. Who was right? The experiments of the time were contradictory and could not yet provide a clear answer to these issues.

In her major work on physics, *Institutions de physique*, du Châtelet dealt in detail with the *vis viva* debate, which was about how best to measure the "force" of a body and to identify possible principles of its transformation and conservation. She found a decisive proof of Leibniz's and her view in an experiment by Willem's Gravesande. In 1722, the Dutchman had done an experiment in which he dropped brass balls of equal mass at different velocities onto a soft clay floor. He later communicated his results to her personally in an exchange of letters. If Newton's equation $E = mv$ were correct, a ball falling twice as fast as another should sink twice as deep. A ball falling three times as fast would sink three times further into the ground. But's Gravesande had measured something else: a brass ball falling twice as fast left an imprint in the clay four times deeper. If it was hurled down three times as fast, it sank nine times further into the clay. These results suggested that the energy went as the *square* of the velocity, in line with the view of Leibnitz and du Châtelet.

As a follower of Newton's Gravesande had not dared to contradict his view. But du Châtelet combined the practical observations of the Dutchman in his own experiments and Leibniz's idea of living forces, and published the correction to Newton's formula in her *Institutions de physique:*

Now, finally, there was a strong justification for considering mv^2 as a fruitful definition of energy.

In total, there are three passages in du Châtelet's work that make her an important pioneer of relativity theory in the twentieth century:

- The formula $E = mv^2$ is a direct precursor of Einstein's $E = mc^2$.
- The statement that there is "no matter without moving force and no moving force without matter" can also be interpreted as an anticipation of Einstein's equivalence of mass and energy.[3]
- Another of Newton's assertions, which he made without empirical proof, was the absoluteness of space and time: he imagined the course of time and the nature of space as divinely predetermined and thus unchangeable. But for du Châtelet and Leibniz, it was clear that Newton's doctrine of the absoluteness of space and time was wrong, as was his conception of a remote effect of gravitation. Two hundred years later, Einstein showed with his theory of relativity that they were right—space and time are *not* absolute.

Another important contribution to modern physics was her idea that the energies known at the time—kinetic energy, positional energy, and the energy of change of form—could be combined in such a way that their sum remains the same. In the mid-eighteenth century, this idea was still in its infancy. The conservation law for the total energy in closed mechanical systems was not finally recognised and formulated until a hundred years later.

*   *   *

Du Châtelet succeeded in refuting the arguments of the Cartesian Dortous de Mairan point by point, and the latter was eventually forced to withdraw from the controversy with his reputation tarnished. The conflict had been energy-sapping for both of them, but it motivated du Châtelet to delve even more deeply into the differences between Leibniz and Newton. This was what gave her the impetus to translate Newton's *Principia* into French. This translation became du Châtelet's most influential work. As before in her treatment of the "Life of Bees", she had the courage to introduce her own thoughts and improvements. For in her opinion, Newton's explanations, as brilliant as his entire theory was, were in some places too brief, somewhat

---

[3] D. Bodanis, *Einstein's Error—The Drama of a Century Genius*, Basic Books (2017).

inaccurate, and sometimes even erroneous. With the utmost rigour, she identified the places where Newton had relied on untested premises—as in his assumption that space and time were absolute—and improved them in her translation. She worked particularly hard on the third part of the *Principia,* "the world system," in which Newton compared his mathematical results with the properties of nature and drew conclusions about practical applications.

Her courage in correcting one of the greatest physicists of all time concerns above all the infinitesimal calculus, which assumed a central importance in Newton's explanations. The fact that Newton's work, which had appeared almost sixty years earlier, had until then only found familiarity and approval among a few continental Europeans was mainly due to the fact that he had developed very unwieldy and sometimes quite abstruse mathematical ways of dealing with infinitesimals and limiting values in the context of Euclidean geometry. The special merit of du Châtelet's work was that she replaced Newton's notations, which only the best mathematicians could follow, with the notation developed by the German polymath Gottfried Wilhelm Leibniz von Leibniz. This notation was common on the continent and far more practicable. At the same time as Newton, the latter had found mathematically equivalent methods of calculation that were so simple that the differential and integral calculus was included in the mathematics lessons of upper secondary schools only a little more than a hundred years later.

By uniting Newton's thoughts with Leibniz's infinitesimal calculus, du Châtelet contributed significantly to the fact that, with only a slight delay, the epochal significance of Newton's work was finally recognised on the continent. The importance of her translation and adaptation of the *Principia* can be seen in the fact that the British scientific community, which long continued to calculate with the ineffective mathematics of its national hero Newton, lost importance after Newton's death and lagged behind the rest of Europe for many decades. It was only when the cumbersome Newtonian integral calculus was finally replaced by Leibniz's that English natural scientists were able to gradually catch up with their colleagues on the continent.

* * *

Du Châtelet's consistent engagement with Newton's work brought her fame and honours. However, she waited in vain all her life for membership of the *Académie des Sciences* in Paris. Only the Academy of Sciences in Bologna registered her as a member in 1746; no other institute in Europe at that time admitted women officially to its ranks. Du Châtelet was very proud of this recognition from Italy. She felt honoured by her nickname "Newton's Apostle", which was common in France, although she also criticised it very clearly.

She was also happy to accept the terms "Emilia Newtonmania" and "Venus Newton".

Her firmness and clear-sightedness had a less positive effect on her private life. For ideological reasons, Voltaire was an unwavering admirer of Newton, whose mechanical laws were grist to his mill of rationality. Through du Châtelet, he had acquired enough mathematical knowledge to understand Newton's physics, to such an extent that in 1738 he was able to publish the comparatively easy-to-read work *Éléments de la philosophie de Newton*. But since he—unlike du Châtelet—could not follow the difficult mathematics in detail, his uncritical acceptance of Newton's ideas was irrational and did not match his usual standards. Voltaire knew about this inconsistency in his otherwise uncompromising thinking, and he also knew that Émilie du Châtelet recognised this contradiction. He could not bear to appear to his beloved as being in any way flawed in his thinking. Although they remained on friendly terms throughout their lives, this became a major reason for the end of Voltaire and du Châtelet as lovers. For 16 years, she had been his equal intellectual companion and even vastly superior to him in mathematical and scientific matters.

From 1745, she worked tirelessly for four years on the translation of Newton's *Principia* and her commentaries. In May 1748, Émilie du Châtelet began an affair with the poet Jean François de Saint-Lambert and became pregnant once again at the age of over forty. She suspected that this late pregnancy might be life-threatening and applied herself all the more restlessly to her work. On 10 September 1749, six days after the birth of her daughter, she finished her manuscript. And on that very day she died of a pulmonary embolism caused by pregnancy and childbirth. She had reached the age of 42. The final edition of her treatment of Newton's *Principia* was published in 1759, ten years after her death—her additions made up a quarter of the two-volume total. To this day, it is the official translation of Newton's main work into French.

Émilie du Châtelet's early death had a direct effect on the Enlightenment in Germany: only now did Voltaire, to whom du Châtelet had felt so attached even after their separation, accept Frederick the Great's invitation to Potsdam. He wrote:

> No woman was ever more learned than she, but no one less deserved to be called a bluestocking.[4] She only ever talked about science with those from whom she thought she could learn something; she never discussed it in order to draw attention to herself (...). For a long time she moved in circles that did

---

[4] *In the eighteenth and nineteenth centuries, a term for an educated woman who put aside her femininity.*

not know her worth, and she paid no attention to this ignorance. (...) One day I saw her dividing a nine-digit number by nine other numbers, in her head, without any help, in the presence of a mathematician who could not keep up with her.[5]

Voltaire's appreciation of his lady friend's scientific achievements was one of the last for a long time. Du Châtelet had corresponded intensively with famous mathematicians such as Johann II Bernoulli and Leonhard Euler. During her lifetime, her works were published in Paris, London, and Amsterdam, and they were reprinted again and again; they were translated into German and Italian and discussed in the most important scientific journals of the era. Yet almost immediately after her death, begrudging contemporaries began to actively denigrate her work. Others simply suppressed her significant contribution to the development of scientific thought in France and her contribution to the development of modern science. The idea that a woman had been instrumental in developing many of the most important ideas in Newtonian physics, which formed the basis of science for the next 150 years, was considered so strange that most publications in the following period concealed her authorship. One example was the *Encyclopédie*, published from 1751 by Denis Diderot and Jean-Baptiste D'Alembert. This was a review of the current state of knowledge at that time, and was considered one of the most important works of the French Enlightenment: a whole series of articles in this standard work was taken from du Châtelet's book *Institutions de physique*, without her being cited as the source.

Despite the systematic and source-critical examination of du Châtelet's enormous achievement that began in the mid-twentieth century,[6] the credit for popularising Newton's physics and helping it to achieve a breakthrough is still largely attributed to Voltaire, even though he did not have anywhere near the mathematical knowledge necessary for this. The fact is that Newton's success on the continent is almost entirely due to Émilie du Châtelet. Her work *Institutions de physique* as well as her translation and commentaries on Newton's *Principia* contributed significantly to the acceptance of Newton's mechanics and thus to the completion of the scientific revolution in France. It

---

[5] Voltaire, *Preface on Marquise du Châtelet*, in I. Newton, *Principes mathématiques de la philosophie naturelle*. Reprint of the French version (1990).

[6] See also J.P. Zinssers (ed.), on du Châtelet, *Selected Philosophical and Scientific Writings*, University of Chicago Press (2009): here especially the section: "Du Châtelet's Contemporary and Subsequent Reputation" (pp. 16–22).

is largely thanks to her that physics and mathematics became the foundations of the Enlightenment.[7]

The world had to wait many hundreds of years for the two great female researchers Laura Bassi and Émilie du Châtelet to revitalise the sciences. But from the beginning of the nineteenth century, women became a permanent fixture in research. The days when the history of science would be determined solely by men was over once and for all.

---

[7] R. Hagengruber and H. Hecht, *Émilie du Châtelet and the German Enlightenment*. Springer Verlag (2019).

5

Sophie Germain (1776–1831): The Greatest Mathematician in France

The nineteenth century was the century of science and rapid technological progress—in 1900, most goods could be produced a hundred times faster than in 1800. Never before had there been so many important physicists, chemists, and mathematicians at work at the same time. But the share of women in these developments was still modest.

Sophie Germain was one of the first women to make her mark on her subject in the nineteenth century. Like Émilie du Châtelet and Laura Bassi, she had had no chance of a regular scientific education at a public institution, let alone at a university. But unlike her predecessors, both of whom had been taught from childhood in their family homes by select teachers, some of them famous, Germain had another hurdle to overcome: she had to acquire her knowledge herself. Her father, a wealthy Parisian merchant, had an extensive library whose precious volumes his three daughters were also allowed to read. As a man of the Enlightenment, he did not stand in the way of his children's thirst for knowledge, but it never occurred to him to hire them a private tutor.

Her father's books opened up a whole new world for Sophie Germain. Even as a small child, she was fascinated by the story of Archimedes' death. The greatest mathematician of antiquity was killed by a Roman soldier in 212 BC during the conquest of his hometown of Syracuse because he preferred to continue drawing geometric figures in the sand rather than escape to safety—his last words, "Do not disturb my circles!", which he shouted to

L. Jaeger, *Women of Genius in Science*, https://doi.org/10.1007/978-3-031-23926-7_5

his murderer, have gone down in history. Sophie Germain marvelled at this devotion to mathematics and wanted to learn more about this science.

Her parents were liberal-minded, but their daughter's early passion for mathematics became sinister to them. Hoping to prevent her nightly reading sessions, they refused to give her the candles she needed and stopped heating her room. It was only when they found their child asleep on a chair in her room one winter morning, huddled under a blanket and frozen to the bone, with a mathematics book beside her, a burnt-down candle she had carefully hidden, and an inkwell whose contents were by then frozen, that they gave up their resistance. From now on, their eccentric daughter was allowed to spend as much time as she wanted doing her studies.

At the age of 13, Sophie Germain began to study higher mathematics, and she was particularly enthusiastic about Leonhard Euler's treatises. At this age, she also learned how to carry out complicated divisions, multiplications, and logarithms using the mathematics textbook *Traité d'Arithmétique* by Étienne Bezout; and this already involved complex numbers, such as the square root of -1. She then taught herself infinitesimal calculus, both in the form of Leibniz's and du Châtelet's modern presentation and in Isaac Newton's more complicated formulation, which had already overwhelmed Voltaire. Germain also acquired Latin through self-study, in order to read Newton's, Leibniz's, and Euler's works.

In 1794, the fifth year of the French Revolution, Sophie Germain turned 18. That year, the *École centrale des travaux public,* which later became the *École Polytechnique,* was founded in Paris. This was a new college set up explicitly to train mathematicians and scientists so that they could contribute to their country, regardless of whether the students were of noble origins or not. How exciting this foundation must have been for the bourgeois Sophie Germain! But even though the motto of the French Revolution was "*Liberté, fraternité, égalité ou la mort*" (Liberty, fraternity, equality or death), one group of people was still denied access to the new academy: women. Fortunately for Germain, the lecture notes of the new university were made accessible to non-students. She thus had the opportunity to learn from many prominent mathematicians of the day. She was particularly interested in the lectures of the great mathematician Joseph-Louis Lagrange on infinitesimal calculus.

* * *

Sophie Germain had managed her studies alone so far; no one in her environment could have given her feedback on her thoughts. Accordingly, she was unsure how to evaluate her knowledge of mathematics. She also longed for a direct exchange with one of her role models. When the student Auguste

LeBlanc, who had provided her with teaching material from Lagrange's lectures, died in the turmoil of the French Revolution, Germain saw an opportunity: she sent her own work to Professor Lagrange under his name. The latter was impressed and soon wanted to meet in person the student who had formulated such astonishing mathematical ideas. Germain summoned up all her courage and turned up for the appointment. Lagrange was blindsided when he saw that "Monsieur LeBlanc" was a woman, but in recognition of her abilities, he became her mentor. Thanks to his support, Germain's name quickly became known and respected in the educated circles of Paris. This success encouraged her to enter into correspondence with other eminent mathematicians. In this correspondence, she initially used her pseudonym "Monsieur LeBlanc", but later she would be able to discuss openly as a woman with almost all the great mathematicians of her time.

Germain was fortunate to witness and be part of one of the most exciting phases in the history of mathematics: the development of number theory. This deals with the properties of the integers, one of the founding disciplines of mathematics and still one of its central themes today, now being systematically explored for the first time as the basis of all mathematics. In 1798, Adrien-Marie Legendre, one of the most famous mathematicians of the day, published his "Essay on the Theory of Numbers" ("*Essai sur la Théorie des Nombres*," or for short, "*Théorie des Nombres*"). This was followed in 1801 by the greatest work ever written on this subject: *Disquisitiones Arithmeticae* by Carl Friedrich Gauss, through which, according to today's understanding, number theory first became an independent scientific discipline. At this point, only a few mathematicians were able to understand Gauss's monumental work—Sophie Germain belonged to this select circle. Thanks to the inspiration of Legendre and Gauss, her own thoughts on number theory reached new heights. She exchanged numerous letters with Legendre. In 1804, she also began corresponding with Carl Friedrich Gauss, whose work on number theory fascinated her. In both cases, she used her pseudonym to conceal her female identity.

Even though Gauss often only replied to Germain's letters after months, he was so impressed by "Monsieur LeBlanc's" work that he praised the "unknown Frenchman" in letters to other mathematicians. Since he hardly ever spoke in praise of professional colleagues, this was a special honour. It was not until 1806 that Gauss found out who Monsieur LeBlanc really was. At that time, the Napoleonic Wars were raging and French soldiers were occupying Gauss' hometown of Brunswick. Sophie Germain feared that her esteemed pen pal might suffer the same fate as Archimedes. So she wrote to General Pernety, a friend of the family, asking him to ensure the safety of the mathematical genius. The latter sent the head of a battalion to Gauss. The concern had been unfounded, but the answer to Gauss' question as to whom he owed this special treatment caused confusion. A Sophie Germain was unknown to him. After some toing and froing, the identity of his supposed correspondent was finally revealed.

Like Lagrange, Gauss reacted to the discovery that "Monsieur LeBlanc" was a woman, not with annoyance and rejection, but with great admiration. He was enthusiastic about the extraordinary talent of his penfriend and the courage with which she had dared to tackle the difficult problems of number theory. In a letter to her in the same year, Gauss wrote:

> But how can I describe my astonishment and admiration when I saw my esteemed correspondent Monsieur LeBlanc converted. (...) But when a woman, because of her gender, and because of our habits and prejudices, encounters infinitely more obstacles than a man, when she makes herself familiar with tricky problems, and yet overcomes these hurdles and manages to reach the greatest depths, she undoubtedly has the noblest courage, an extraordinary talent, and a superior genius.

Unfortunately, the exchange on number theory between Sophie Germain and Carl Friedrich Gauss ended shortly afterwards. Germain described some of her further ideas on this area of mathematics in a letter in 1808, but she never received a reply. Gauss had accepted a position as professor of astronomy at the University of Göttingen and ceased his investigations of number theory. Germain now also turned to another subject.

* * *

In 1808, the German physicist and astronomer Ernst Chladni showed a fascinated audience of Parisian scholars how a violin bow set in vibration created visible patterns in a layer of sand that had been spread out on a flat glass plate. Only certain areas of the plate vibrated and the sand accumulated where the plate was motionless. It had long been known that musical notes and sounds were caused by vibrations, and the vibrations of stringed instruments in particular could be calculated with the help of differential equations. Leonhard Euler had already described the mathematics of a vibrating bar using this mathematical tool. So in the one-dimensional case, the matter was clear. But the experiment with the violin bow and the plate sprinkled with sand moved into far more demanding areas of mathematics. How could oscillating behaviour in *two-dimensional* space be handled mathematically?

Chladni's experiment challenged scholars. In 1809, Emperor Napoleon himself offered a prize for the development of a mathematical theory of elastic surfaces. This prize also aroused Germain's interest, because she was familiar with differential equations. But as it turned out, vibrations in two-dimensional space require a far more complicated analysis than vibrations in one-dimensional space. Due to the additional dimension, the number of possible convergence patterns virtually explodes, and with it the complexity of the required differential equations.

Germain's friend and patron Lagrange advised her and other mathematicians against trying to solve this problem. He knew that this would require a whole new system of analysis that a single mathematician could not possibly develop on their own. But Germain was not discouraged. Her approach was to apply a new method of Lagrange's, the so-called calculus of variations, to Euler's earlier results. Her idea was a good one, but now the failings of her previous life came to bear. She had never received a formal education, and her mathematical knowledge was correspondingly fragmented. While in some areas she was able to exchange ideas with the best mathematicians to mutual benefit, sometimes even surpassing them, there were clear gaps in other areas. Moreover, social conventions made it difficult for her, as a single woman, to communicate with other mathematicians in an unbiased manner. The correspondence shows, for example, that most correspondents had inhibitions about hurting a young woman's feelings by criticising her work.

Germain did not perfectly master the calculus of variations, so errors crept into her calculations. With the help of her mentor Legendre, who nevertheless found her ideas impressive, she worked out a solution and submitted it in September 1811. She was the only participant. No other mathematician had found an even moderately passable way of doing the calculation. However, the jury decided that Germain's contribution did not meet the requirements and extended the competition for another two years.

Germain tried again and in 1813 received no more than an honourable mention for her next contribution, which again contained some errors. That year, her fatherly friend Lagrange died, so she had to continue working without his help. But Germain did not give up. In the meantime, a competitor had come forward: Siméon Denis Poisson, a 33-year-old physicist and mathematician from Lagrange's circle who had quickly made a career in academia. In August 1814, he claimed to have found the solution and tried to have the new edition of the prize suspended so that he could receive it himself later. But it quickly became apparent that his formulas were no more than a rehash of Germain's second attempt and also contained errors. So the race was open again.

In 1815, Sophie Germain submitted a third paper on the vibrations of elastic surfaces. The paper *"Recherches sur la théorie de surfaces élastiques"* (Research on the Theory of Elastic Surfaces) contained the essence of her years of research; the manuscript had only half the scope of her earlier attempts, but it contained significant new calculation methods. Her solutions even went beyond the required problem: she did not limit herself to the investigation of strictly two-dimensional planes, but allowed for curved surfaces. With this extension of the work previously submitted by herself and by Poisson, she once again clearly set herself apart from her competitor. The latter had not entered any new work after his attempt of 1814.

After seven years of tireless work, the now 39-year-old Sophie Germain finally became a laureate of the Paris *Académie des Sciences on* 8 January 1816, an honour that Émilie du Châtelet and Voltaire had failed to achieve eighty years earlier. The prestigious prizes had been offered since 1720, and Sophie Germain was the first woman to win one of them. But there was a bitter aftertaste mixed in with the triumph. Her competitor Poisson proved to be a bad loser and did not appear at the award ceremony, and like him, some other mathematicians did not recognise the award. Even the judges, although they praised the high quality of the work, could not refrain from pointing out some shortcomings in Germain's analysis. It would be decades before mathematicians of a new generation managed to remedy the remaining short-comings. The greatest humiliation, however, and the height of absurdity, was that Sophie Germain was not allowed to attend her own prize-giving cere-mony, because only women who were wives of members were admitted to the Academy's meetings. It was not until seven years later that she became the first woman to be granted the right to attend the meetings of the *Académie*. And she never obtained a permanent position.

After her triumph, Sophie continued her work on the theory of elas-ticity and wrote several papers on the phenomenon of oscillating planes. In the following years, papers were published by several other mathemati-cians. Among them were Augustin-Louis Cauchy, known to mathematicians for a certain kind of convergent sequence of numbers named after him, and Henri Navier, who later developed new approaches to solving the Navier–Stokes equation for fluids. The latter made it known with all due fairness that Germain's work had greatly inspired his own. Siméon Denis Poisson, on the other hand, who also further developed the mathematics of oscillating planes and fluid dynamics, concealed the fundamental role of Sophie Germain in his publications. In 1821, Germain's manuscript on the elasticity of surfaces was still unpublished. In order not to leave the field to others and be forgotten, she had to publish her work herself.

Today, Germain's work on vibrating planes is of fundamental relevance to physics and engineering. From the vibrational behaviour of wings in aircraft construction to the sound pressure of loudspeaker membranes, many phenomena can be described and calculated using the corresponding differential equations. Germain's greatest achievement, however, concerns a completely different field of mathematics.

* * *

In 1637, some 180 years earlier, the great and rather enigmatic mathematician Pierre de Fermat had come up with an equation that he had jotted down in the margin of a page of the ancient textbook *Arithmetica* by Diophantus of Alexandria:

$$x^n + y^n = z^n$$

For this equation, n = 2 gives the Pythagorean theorem, which has infinitely many positive integer solutions. The triplets x = 3, y = 4, z = 5 (9 + 16 = 25) and x = 10, y = 24, z = 26 (100 + 576 = 676) are just two examples from the infinite set of possible triplets. Fermat claimed in his marginal note that, for n > 2, this equation has no solutions for positive integer x, y, and z. He wrote:

> It is not possible, however, to decompose a cube into 2 cubes, or a biquadrate into 2 biquadrates, and in general a power, higher than the second, into 2 powers with the same exponent: I have discovered a truly marvellous proof of this, but this margin is too narrow to contain it.[1]

Fermat's conjecture that $x^n + y^n = z^n$ for natural numbers (positive integers) x, y, and z only has solutions when n = 2 is known today as "Fermat's last theorem". Fermat's solution was not found anywhere in his papers. In the following centuries, the search for this "truly miraculous" proof drove many a mathematician almost mad. It was as if Fermat had left a treasure map, only with no clue as to where in the world the treasure was actually located. Mathematicians can live with things if a certain proof cannot be carried out in principle. But knowing that there was apparently a proof that could almost

---

[1] Unfortunately, Fermat's original writing has been lost. But one finds his conjecture repeated in numerous works, most prominently in the edition *Arithmetica* by Diophantus with notes, which Fermat's son edited, see P. Tannery and Ch. Henry (eds.), *Œuvres de Fermat. Tome Premier*. Paris: Gauthier-Villars (1891), p. 291; see also: https://gallica.bnf.fr/ark:/12148/bpt6k6213144d.texteImage).

fit on the margin of a page, but not being able to figure out how it worked, was a never-ending torment for them.[2]

About a century later, Leonhard Euler provided a proof using complex numbers that Fermat's conjecture is valid for n = 3 and also for n = 4. Later, this proof turned out to be incomplete, but Carl Friedrich Gauss succeeded in completing it. For n > 4, however, his arguments no longer applied. It was also shown that the theorem only had to be proved for values of n that were prime numbers. For if n can be decomposed into multiplicands, it is sufficient to prove Fermat's conjecture for only one of these multiplicands. Gauss had thus proved that Fermat's conjecture holds for all n that are multiples of 3 or 4.

This was the state of research when Fermat's conjecture aroused Sophie Germain's interest in number theory. In 1819, after a break of almost ten years in her correspondence, she again contacted her great role model in this field, Carl Friedrich Gauss. She wrote to him that she had succeeded in showing, using a method all her own, that there were no solutions to Fermat's conjecture for n = 5 unless x, y, and z were enormously large numbers. Self-critically, she added that "enormous" was of course not a suitable mathematical category; a mathematical proof had to apply to all numbers. Later, in an unpublished manuscript entitled *Remarque sur l'impossibilité de satisfaire en nombres entiers a l'équation $x^p + y^p = z^p$*, Germain showed that, for p > 5, all potential counterexamples to Fermat's theorem must be numbers so large that their "size terrifies the imagination"[3]—her calculations had entered a realm where numbers consist of about forty digits (i.e., ca. 10^{40}).

In her letter to Gauss, Germain also outlined a strategy for a general proof of Fermat's last theorem. The letter contained substantial progress towards a proof of special cases, in particular an important special case that she was later able to actually solve. When she asked whether it was worth pursuing her approach further, she received no answer from Gauss; he had finished with number theory.

Even without the support of the mathematical genius, Germain managed to make significant contributions to the proof of Fermat's conjecture. Crucially, she saw little point in trying to tackle the problem for individual numbers, as many mathematicians had been doing. Gradually, they had succeeded in confirming Fermat's conjecture for specific numbers. For the

---

[2] Today it is assumed that Fermat had not found a proof for all values of n, but "only" for n = 4.

[3] A. Del Centina, *Unpublished Manuscripts of Sophie Germain and a Revaluation of Her Work on Fermat's Last Theorem"*. *Archives for History of Exact Sciences*, 62, 4 (2008), pp. 349–392; https://link.springer.com/article/10.1007/s00407-007-0016-4.

number 5, Legendre gave the proof in 1825—followed almost simultaneously by the German mathematician Johann Peter Gustav Lejeune Dirichlet.
Shortly afterwards, the matter was also clear for n = 7, 11, and 13. But
then the search stalled due to the complexity of the proofs involved. It
was not until 1943 that the 19-year-old Canadian mathematician James Fell
succeeded in proving the conjecture for n = 17 and n = 23.[4]

Germain took a different approach. She saw that the only fruitful approach
was to prove Fermat's conjecture for whole classes of numbers. In fact,
between 1819 and 1821, she provided a proof that Fermat's conjecture is
correct for all prime numbers p for which 2p + 1 is also a prime number—
this is "Sophie Germain's theorem," named after her today. The number 5
is an example of such a prime number, because 2 * 5 + 1 = 11, and 11
is another prime number. Primes that satisfy this condition are today called
"Sophie Germain primes" in her honour.

Since Germain did not publish this proof, nor her other contributions
to the proof of Fermat's conjecture, her rather spectacular theorem initially
found little circulation. Legendre was the only one who explicitly referred to
Germain's proof of prime numbers for Fermat's conjecture on a few occasions.

> This demonstration, which will doubtless be thought very ingenious, was made
> by Miss Sophie Germain, who successfully cultivates the physical and math-
> ematical sciences, as is proved by the prize she won at the Academy on the
> vibrations of elastic plates (sur les vibrations des lames élastiques).[5]

Both Legendre and Dirichlet used Germain's theorem, which provides the
necessary foundations to fully verify Fermat's conjecture for n = 5.

Recent studies of some of Sophie Germain's unpublished manuscripts and
letters show that her theorem on special primes was only one of many steps in
her grand plan to prove Fermat's conjecture in its entirety.[6] She was the first
person to define a coherent plan for proving Fermat's conjecture. She worked
tirelessly for years on this plan with the help of her modular arithmetic, but
it was to take another 175 years before it could be completed.

* * *

---

[4] *Elementary proofs of Fermat's great theorem for some special cases.* Dtsch. Math. 7 (1943).

[5] Adrien-Marie Legendre, *Recherches sur quelques objets d'analyse indéterminée et particulièrement sur le théorème de Fermat*, in *Mémoires de l'Académie royale des sciences de l'Institut de France*, 6 (1823).

[6] A summary of her actual achievements can be found in: H. Kagele, *Sophie Germain, The Princess of Mathematics and Fermat's Last Theorem*, https://www.gcsu.edu/sites/files/page-assets/node-808/attachments/kagele.pdf.

Sophie Germain's discovery of the primes now named after her was the greatest advance on the way to the final proof of Fermat's conjecture. For the first time, it had been possible to prove this conjecture for a very large, presumably even infinitely large number of exponents (whether the set of Sophie Germain primes is infinitely large has not yet been conclusively answered), and not just for specific natural numbers. Unfortunately, many primes do not fall into the class of Germain's primes, among them are, for example, 17 and 19. (In 1915, it was proved that there are infinitely many primes that are not Sophie-Germain primes.)

Another important contribution made by Germain was to use her own theorem to check Fermat's last theorem by hand for $p < 100$ in the case where x, y, and z are prime numbers, so that there was a limited set of equations to work through. Legendre repeated and checked her calculations. In 1920, almost a hundred years later, the American mathematician Leonard Dickson used Germain's theorem to prove Fermat's conjecture for $p < 1700$.

Mathematicians continued to work on this topic for many, many years, and Germain's findings were an important basis for all of them. It was not until 1995, more than 350 years after Fermat had written his momentous remark on a page margin, that the British mathematician Andrew Wiles was able to prove that Fermat's conjecture is indeed correct for all natural numbers. Thus, Fermat's conjecture finally became Fermat's theorem, because in mathematics one only speaks of a "theorem" when there is a proof for it. The final proof of Fermat's theorem was made possible using the highest level mathematics, which was still completely unknown to the mathematicians of the nineteenth century, and indeed to Sophie Germain when she did her preliminary work.

* * *

Due to the still massive prejudices against women, Sophie Germain was denied an academic career. Even though she worked on number theory with Legendre, and indeed on equal terms with the great mathematician, in the 1820s, she was never offered a position at a university. Throughout her life, she worked in private. She initially received the financial means to devoted her life to mathematics from her father. When he died in 1821, she was supported by her sisters, who had married wealthy men. Her family refrained from pressuring her to marry a well-to-do man, a very tolerant attitude in those days. Nothing is known about any lovers she may have had, and little is known about her life between 1826 and 1829. In 1829, a student of the German mathematician Bader travelled to Paris and reported to her that Gauss had published his *Disquitiones generales circa superficies curvas*

("General Investigations on Curved Surfaces") in 1827. In this work he had formulated his *Theorema egregium* ("Remarkable Theorem"). This offered a more general solution to the problem of curvatures that had occupied Germain for many years. So Sophie Germain once again took up contact with Gauss:

> I regret that I cannot submit to your judgment a variety of ideas which I have not published and which would take too much time to write down.

In the same year, Germain discovered that she had breast cancer, from which she died two years later, on 27 June 1831. At this time, the great Gauss had advocated to the University of Göttingen that she be awarded an honorary doctorate and a professorship. But this advocacy came too late. On her death certificate, Germain is described as a pensioner and building owner. The fact that she was the most important mathematician of her time was not taken into account, probably because she had never enjoyed a formal education. Even today, her importance for mathematics tends to go unnoticed: among the names of the 72 most important scientists and engineers of the eighteen and nineteenth centuries, which are inscribed in gold letters on the outside of the first floor of the Eiffel Tower, one looks in vain for her name. And no other woman has been included in this list.

In mathematical circles, on the other hand, she has not been forgotten, and not only because a mathematical theorem and a class of prime numbers have been named after her. The greatest mathematician of the day, Gauss himself, paid tribute to her in 1837, six years after her death:

> She has proved to the world that even a woman can achieve something worthwhile in the most rigorous and abstract sciences, and therefore deserves an honorary title.[7]

---

[7] N. Mackinnon, Sophie Germain, or, Was Gauss a Feminist?, *The Mathematical Gazette*, 74, 470 (1990), pp. 346–351; https://www.cambridge.org/core/journals/mathematical-gazette/article/abs/sophie-germain-or-was-gauss-a-feminist/6176F6C98067333F574636CD4A40D22C.

6

Caroline Herschel (1750–1848): The Great Astronomer in Her Brother's Shadow

Caroline Herschel was born almost a quarter of a century before Sophie Germain, who helped to shape mathematics at the end of the eighteenth century and the beginning of the 19th. The fact that Herschel is considered here as a scientist of the nineteenth century is due to her advanced age: she died in 1848 at the age of 97.

Her brother Wilhelm Herschel had moved from Hanover to England and later became world famous as the discoverer of the planet Uranus and explorer of the night sky. The nearly seventy treatises he had published under his name with the *Royal Society* were significant contributions to celestial science and cemented his status as one of the greatest astronomers of all time. The fact that Caroline Herschel, who was twelve years younger and whom Wilhelm had brought to England to live with him, was significantly involved in this work remained unknown for a long time. She later wrote in her memoirs:

If I said that I had published all my brother's works, no one would believe it

Although she was expected to behave in a humble and reserved manner from an early age, she managed to step out of her beloved brother's shadow. Over time, she outgrew her original role as a housekeeper and assistant and secured the respect of her contemporaries with her own observations of the heavens. A true pioneer, she was the first woman to be employed at the Royal Observatory in Greenwich, making her the first female astronomer of the modern era to be awarded a salary for her work.

© The Author(s), under exclusive license to Springer Nature
Switzerland AG 2023
L. Jaeger, *Women of Genius in Science*,
https://doi.org/10.1007/978-3-031-23926-7_6

The esteem in which she was held was also unprecedented. No woman before her had ever been awarded a gold medal by the *Royal Astronomical Society*, let alone made an honorary member. Even at an advanced age, Caroline Herschel was showered with honours. In 1835, when she was in her mid-eighties, she became the first woman to be made an honorary member of the *Royal Astronomical Society*. Three years later, the Royal Irish Academy of Sciences accepted the now 88-year-old into its ranks, and in 1846 she received the Gold Medal of the Prussian Academy of Sciences on behalf of the King of Prussia. The great naturalist Alexander von Humboldt did not miss the opportunity to personally present the 96-year-old with this extraordinary award.

* * *

Caroline Herschel was the eighth of ten children of the Hanoverian military musician Issak Herschel and his wife Anna Ilse Moritzen. Four of her siblings died in infancy. Caroline's life was also in danger. At the age of three, she contracted smallpox; at that time, one in ten children died of this disease. At eleven, she contracted another of the dreaded infections of the day: typhus, also called typhus fever. She survived this disease too, but her body was so weakened that she stopped growing. As a result, she was a strikingly short person as an adult, measuring only about 140 cm. Her short height, the lack of a trousseau, and the ugly scars left by the smallpox destroyed any hope of establishing a household of her own through marriage. The way things worked in those days, this would mark out her future: Caroline Herschel would be available to her parents and siblings all her life as a domestic servant, contributing to the family coffers through needlework. In the opinion of her mother, who could hardly read and write herself, an education beyond the bare minimum was not necessary for this path in life. Fortunately for Caroline, her father disagreed.

Isaak Herschel was ambitious and musically gifted. His family had lacked the means for his education, but he had managed to work his way up under the most difficult conditions. He taught himself to play the oboe and found a permanent position as a musician in the military. This experience may have contributed to the special importance he attached to a good education for his children. He even allowed his daughter Caroline to attend the garrison school together with her brothers. All of his sons later made a living as musicians; the eldest daughter, who tended to follow her mother, married a violinist. So music was omnipresent in the Herschel household. But they were also extremely interested in philosophical questions—and in particular, astronomy. Herschel senior often showed his children the stars and planets in the night sky, and explained their motions to them. Special lessons in mathematics and French were organised for the sons, from which Caroline benefited indirectly. For what they had learned was discussed in the evenings, much to the displeasure of their mother, who did not fully appreciate Caroline's thirst for knowledge being assuaged in this way. Thus Caroline's father provided a ray of hope in her life, the hope that she might be able to escape the narrow confines generally envisaged for women in those days.

* * *

Caroline's father had suffered ill health from his life as a soldier, and in 1767, when she was 17 years old, her father died. Her dreams of an interesting future suddenly became a thing of the past. Now her mother and her eldest brother Jacob, who also believed that a woman's place was in the home, were left in charge. The next five years must have been difficult for Caroline, because all she had to do now was housework. Her favourite brother Wilhelm had left for England many years before. The few opportunities she had to learn anything were secret reading sessions and the occasional opportunity to sing arias. Later she wrote about her mother in her memoirs:

She had decided that I was and should remain a brute, though a useful one

In 1772, there was another decisive turn in her life. Caroline's brother Alexander mentioned in a letter to Wilhelm, who had established himself as a musician, music teacher, and composer in the elegant English spa town of Bath, that Caroline was talented at playing the violin and had a beautiful singing voice. Wilhelm came to Hanover for a visit and was soon convinced of his 22-year-old sister's talent. He suggested that he take her to Bath and train her as a singer. Caroline's mother and her eldest brother Jacob, who was head of the family after their father's death, had little sympathy for this plan. They only let Caroline go when Wilhelm had agreed to provide money to hire a new domestic helper. Later, Caroline told Herschel that she had knitted stockings for the family for two years during the protracted negotiations about her future fate.

For Caroline, this deal was a genuine liberation: she was to go with her beloved brother to a foreign country whose language she did not know. In Bath she lived in the house of her brother, here called William, and received vocal training. Her debut in 1777 as the lead voice in an oratorio by George Frideric Handel was a success, and further performances followed. However, she turned down the offer to move to Birmingham, 175 kms away, to devote herself entirely to her career as a singer. Instead, she continued to perform in Bath and the surrounding area and kept house for her brother.

* * *

Like his father Isaak Herschel, the respected musician and composer William Herschel also developed another passion: for astronomy. At first, his sister was not so fond of this enthusiasm; keeping house in addition to her rehearsals and performances as a singer already meant a multiple burden. The more obsessively William turned to the stars, the more trouble it meant for Caroline—sometimes she had to feed her brother with snacks during his nightly observations. But often, she would sit in the house and note down the observations as William called them out to her through the window. If William had had to write down his observations himself, he would have had to expose his eyes to the candlelight again and again, and a great deal of time would have been lost until they got used to the darkness again.

For Caroline Herschel, this auxiliary service meant that, like her brother, she hardly slept. But she also benefited from the collaboration, because William taught her the mathematics needed for the complicated position determinations. She wrote about this time in her memoirs:

> I was a mere tool which, for want of a better one, he took pains to sharpen and make fit for his purposes

In 1774, William Herschel began to build his own telescopes, because the ones available on the market either had too low a magnification for his requirements, or they were inaccurate or too expensive. The construction of telescopes in which glass lenses were arranged one behind the other was out of the question for various reasons. Among other things, the glass lenses would have had to be far too large and heavy for the unprecedented magnifications he wanted to achieve, given the state of knowledge at the time. And since the light had to pass through them unhindered, they could not be supported on mounts either. William thus decided to build reflecting telescopes; with this kind of device, a concave metal mirror attached to the lower end of the telescope tube collects the incident light and, to put it simply, reflects it into the observer's eye. The manufacture of the mirrors required a great deal of patience and concentration, because their surface had to be extremely smooth, and it also had to be machined with absolute precision to the calculated curvature. To exclude optical distortion, the deviation from the ideal shape could not be more than one eighth of the observed wavelength. For visible light, that meant an order of magnitude of 50 nm, or 0.00005 mm.

As a devoted sister, Caroline not only allowed the house she ran to be transformed into a workshop in which metal alloys were melted and cast, she also got involved in the enormously time-consuming and tedious work of mirror polishing. It was indeed virtually endless, because as the metal tarnished over time, the mirrors had to be repeatedly removed from the telescopes, polished, and reinstalled.

This smooth collaboration made Caroline and William Herschel one of the most effective teams science had ever witnessed. Caroline recorded the observations, made the necessary calculations to standardise them with factors such as the time in question, and then compiled the data for publication in scientific journals. In time, William Herschel gained a good reputation among astronomers. Caroline and another brother, Alexander, who was also an indispensable help in grinding the mirrors, remained in the background.

* * *

In 1781 came the great breakthrough: William discovered a new planet during his survey of the night sky! The planets visible to the naked eye— Mercury, Venus, Mars, Jupiter, and Saturn—had been known since time immemorial. Together with the Sun and Moon, the number of moving celestial bodies was seven, a magical number in many cultures. The seven celestial bodies also explain the number of days in the week, which in many cultures is seven: each of them is assigned to a celestial body or the god it symbolises. The discovery of a new planet was a unique moment in human history, because it broke the supposedly divine pattern. Now astronomers were asking themselves: why had a planet, visible to the naked eye under the most favourable conditions, remained undiscovered? And if a planet had remained undiscovered in this way, what else could be out there?

Later it turned out that several astronomers had previously discovered and catalogued the same spot of light at various points in the firmament, but had not recognised it as a planet. Even William Herschel, who had looked more closely and patiently than anyone before him, did not initially think of the new object as a new planet, even though it had noticeably changed its position in a short time. However, he soon had to revise his assumption that it must be a comet. He was well connected in astronomical circles, and asked some of his contacts for advice. The Astronomer Royal Nevil Maskelyne examined his discovery and confirmed that it must be a previously unknown planet. Other colleagues were sceptical: could a layman have discovered something so significant? A few months after Herschel's discovery, the Scandinavian astronomer Anders Johan Lexell had calculated the object's orbit beyond any reasonable doubt as being approximately circular around the Sun, thus proving that it had to be a planet. William Herschel thus became a hero of the scientific world.

King George III, himself an enthusiastic amateur astronomer with his own observatory, sent for William Herschel in May 1782. After some shop talk, he was soon convinced of his counterpart's knowledge and appointed him as the Astronomer Royal. Herschel's decision to name his new planet "George's Star"—the name Uranus only became established later—may have contributed to this appointment. With his new position as court astronomer and the annual salary of 200 pounds granted by the king—about 45 000 pounds in today's values—William Herschel could now devote himself entirely to astronomy.

* * *

William's success had a direct impact on his sister's life. In his new position, he needed her support around the clock, so he asked her to give up her profession as a singer. In fact, she followed him from the fashionable city of Bath, where England's upper classes regularly gathered, to a small village near Windsor. Here, conditions were more favourable for nightly stargazing, and the members of the royal family who came to stay with the Herschels did not have to travel a long distance. Caroline Herschel was now 32 years old and once again her own interests had taken a back seat. But instead of capitulating to a life that did not favour women, she took things into her own hands. With increasing enthusiasm, she began quite independently to scan the firmament for celestial objects.

Shortly before this, the French astronomer Charles Messier had published a catalogue of 103 mysterious celestial bodies that were visibly extended in the field of view of the telescope; that is, they had spatial extension on the celestial sphere, in contrast to the stars, which are point-like even at high magnifications. The siblings became interested in these irregularly shaped objects called stellar nebulae, and they were the first to discover that some of them are unusually closely spaced fixed stars, so-called star clusters. In 1786, Wilhelm Herschel submitted to the *Royal Society of London* a list of stellar nebulae that he and Caroline had found during their systematic exploration of the firmament. Over the years, they added to this compilation several times, and literally thousands of celestial bodies were eventually located. Today, the stellar nebulae are identified as star clusters, interstellar galactic nebulae, and galaxies.

William increasingly perceived his sister as an independent researcher and left her a new and better observing instrument. The court astronomer Nevil Maskelyne also showed appreciation for her work and completed her training as an astronomer. And yet her work was still known only to a small circle of experts.

* * *

In 1786, Caroline Herschel made a discovery that also brought her the attention of the general public. William was travelling for a few weeks in summer on behalf of the King, and Caroline was able to use his telescope. On 1 August, Caroline achieved something that had always been a holy grail for astronomers: she discovered a new comet. One day later, she wrote a letter to the secretary of the Royal Society in which she reported her find, beginning with the words:

> In consequence of the friendship that exists between you and my brother, I take the liberty of bothering you with the following imperfect report

After the announcement of their discovery, countless congratulatory letters arrived, and their teacher Nevil Maskelyne enthusiastically described their observation as a "triumph for British astronomy". Prominent visitors came to the Herschels' house. Many women were delighted that a female colleague had made such a significant discovery. It was also a great success that one year later Caroline Herschel's letter was published in the *Philosophical Transactions*, the most influential journal of the *Royal Society of London* at the time. She thus became the first woman in England whose scientific work on astronomy was published under her own name. In 1787, King George III again showed his generosity by granting Caroline Herschel a position for life, with an annual salary of 50 pounds. And once again Caroline Herschel had broken through a glass wall: with this appointment, she became the first woman in England to receive a regular salary as an astronomer. She had made it: instead of having to hold embroidery up close to her eyes, as her mother had planned for her, her gaze would now extend into the endless expanses of space.

Inspired by this recognition, Caroline Herschel began to search the firmament for more comets with ever better constructed telescopes. As early as 1788, she spotted a second specimen, but a competitor narrowly beat her to it. In December of the same year, she spotted her third comet. Further comet sightings followed in December 1791, October 1793, November 1795, and August 1797. In all, Caroline Herschel found nine comets, five of which she was undoubtedly the first to discover and describe. An anecdote describes how the generally docile and humble Caroline Herschel could also be very assertive: the eighth comet she discovered was comparatively bright, so it was likely that other astronomers would notice it too. In order that no competitor could dispute her priority, she rode 30 miles to Greenwich the following day—something unheard-of for a woman—and made the position of the new comet public there.

Caroline Herschel's comet discoveries earned her growing recognition, and not only among experts. Her fame was also so great among the educated middle classes that everyone knew who was meant in the caricature "The female philosopher: smelling out the comet".[1] It shows a woman kneeling in front of a telescope, enraptured by the sulphurous yellow ray that streams out of the backside of the figure representing the comet in the sky above her.

Although it was customary to name a comet after its discoverer, none of the comets first described by Caroline Herschel were named after her during her own lifetime. The comet she discovered in 1788 was rediscovered by the French astronomer Roger Rigollet in 1939 and has since been named 35P/Herschel-Rigollet. It is expected to be visible from Earth again in 2092.

Caroline Herschel was not only on the lookout for comets. Three quarters of a century earlier, in 1712, the star directory of the English astronomer John Flamsteed had been published, in which he had catalogued about 2800 stars with their exact positions. After Flamsteed's death in 1719, his work had been supplemented several times, among others by Edmond Halley. In her systematic explorations of the heavens, Caroline Herschel found a total of 561 stars that had eluded her predecessors. In 1798, she submitted to the Royal Society a supplementary catalogue to the *British Catalogue*, which had not yet been published at the time. Attached was a list of errors that had remained in the compilation until then.[2]

At the same time as these successes, a painfully felt change was to occur. In May 1788, her brother, now 50, married the widow Mary Pitt. The couple moved to nearby Upton, leaving Caroline behind. The loss of her role as a housekeeper watching over her brother's welfare must have been a blow to her, even though she now had more time for her studies. But Mary Pitt also had to swallow a very bitter pill: on starry nights, William rushed back to Caroline and his astronomical gadgets. All in all, the newly-wed was not

---

[1] *The Female Philosopher: Smelling out the Comet*, 1790s; Draper Hill Collection, The Ohio State University Billy Ireland Cartoon Library & Museum. See also: https://www.researchgate.net/figure/ The-Female-Philosopher-Smelling-out-the-Comet-1790s-Draper-Hill-Collection-The-Ohio_fig2_273 323994.

[2] C. Herschel, *Catalogue of stars taken from Mr. Flamsteed's observations contained in the second volume of the* Historia cœlestis, *and not inserted in the British Catalogue. With an index, to Point out Every Observation in that Volume Belonging to the Stars of the British Catalogue. To which is added, a collection of errata that should be noticed in the same volume.* Published by Order, and at the Expense, of the Royal Society, London (1798).

so fond of this arrangement and it was not until many years later that the sisters-in-law would eventually strike up a friendly relationship.

* * *

After many years of fruitful collaboration, William died in 1822 at the age of 84. Although Caroline Herschel had spent fifty years in England, she no longer felt any great attachment there. A few weeks after his death, she moved back to her home town of Hanover. Her reputation had preceded her, so that even there the most eminent scholars sought her out in her simple house in Marktstraße, where she lived with her younger brother Dietrich and his family.

Even in her seventies, she remained active in astronomical research. One of her major contributions at that time was to edit the star atlas, as it stood at the time, whose beginnings went back to the aforementioned Flamsteed. Despite her advanced age, Caroline Herschel was chosen to check this standard work, by now known as the *Historia Coelestis Britannica*. With enormous effort, she went through the information given in the celestial atlas, correcting the errors and integrating the 561 celestial bodies she had herself discovered and described in 1798. Caroline Herschel received much recognition for this extensive work, even gaining the admiration of Carl Friedrich Gauss, who was one of the many famous visitors to her house in Hanover.[3]

Another work from this period was a family affair. William Herschel's only son, John Frederick William Herschel, had followed in the footsteps of his father and aunt and had also become a great astronomer—his importance is attested by the fact that, after his death in 1871, he was given a state funeral in Westminster Abbey and laid to rest next to the graves of Charles Darwin and Sir Isaac Newton. Caroline Herschel had a very close relationship with this nephew. When John intended to update the catalogue of stellar nebulae that she and her brother William had once compiled, she promised him her support. In the new catalogue, the stellar nebulae would no longer be arranged according to the classification system she and William had devised, but according to their celestial positions. Once again, Caroline Herschel demonstrated extreme precision, patience, and dedication. For her efficient cataloguing of thousands of stellar nebulae, the *Royal Astronomical Society* awarded her a gold medal in 1828, when Caroline Herschel was 77 years old.

The Herschels' work became the basis of today's *New General Catalogue of Nebulae and Clusters of Stars* (NGC), which contains all known galactic

---

[3] Letter from Carl Friedrich Gauss to Caroline Herschel 1825.

nebulae, star clusters, and galaxies—the so-called deep-sky objects. As already mentioned, William Herschel had published the first precursor in 1786. Even then, Caroline Herschel had played a leading role, although her name was not mentioned. The siblings added to the list several times and recorded and described a total of around 2500 objects, each with meticulously calculated position. Caroline's nephew John Herschel added another 1700 nebulae to the catalogue by the end of the nineteenth century. More than half of the 7600 deep-sky objects known today were thus found and catalogued by a member of the Herschel family.

Even in her eighties, Caroline was still very busy. Her nephew John wrote to his wife in 1832:

> She walks about town with me, skipping up her two flights of stairs, as wonderfully fresh as some people I could name who are not a quarter as old as she is [...] In the morning till eleven or twelve she is dull and tired, but as the day progresses she becomes more and more lively, and at ten or eleven o'clock at night she is quite fresh and merry, and sings old rhymes, nay, even dances to the great delight of all who see her

This was followed by the other honours mentioned at the beginning of this chapter. Caroline Herschel remained mentally active into old age. Already in England, she was used to direct contacts with members of the royal family; King George III, who was enthusiastic about astronomy, had often been a guest of Caroline and William Herschel and had also received them in nearby Windsor. In 1837 Victoria became queen, but in Hanover the law was such that only a male successor could ascend the throne. Hanover seceded from England and one of King George III's sons was proclaimed king there. Caroline Herschel was also close to this line:

> She was in her 96th year when the Crown Prince of Hanover was born, and she expressed her regret that she, who had known personally all contemporary members of the British Hanoverian royal family, would never see this young prince, as she was too weak to leave her home. Upon hearing this, King Ernst-August immediately gave orders for his grandson to be brought to her house, so that this last wish of the revered old lady might also be fulfilled [4]

On her 97th birthday, the Crown Prince and Crown Princess did her the honour of coming to her house for a visit lasting several hours. The atmosphere seems to have been relaxed, as it is said that Caroline Herschel sang

---

[4] Johann Heinrich Mädler, History of Celestial Science (1873).

a song to her esteemed visitors that Wilhelm Herschel had composed many decades before—a reminiscence of the siblings' earlier life dedicated to music.

Caroline Herschel died on 9 January 1848, two months before her 98th birthday. She herself wrote the inscription to be left on her gravestone in Hanover. It begins:

The gaze of the transfigured was turned towards the starry heavens—her discoveries of comets and her participation in the immortal work of her brother, Wilhelm Herschel, will bear for all posterity. [...]

Caroline Herschel is one of the very few women of past centuries whose work was recognised and appreciated during her lifetime. Even though she remains to some extent overshadowed by her elder brother, she has won a place among the great scientists.

Even today, the English Royal Astronomical Society and the German Astronomical Society are closely linked. In 2021, the two associations jointly endowed a medal to be awarded to outstanding female astronomers, alternately from Great Britain and Germany. The prize money of £10 000 associated with this honour is to be used directly for research or for costs associated with such work, explicitly including childcare. And the name of this award is the Caroline Herschel Medal.

7

Ada Lovelace (1815–1852): Inventor of Computer Algorithms

Since ancient times, people have been striving to simplify and speed up arithmetical calculation. As shown by the cargo of a Roman shipwreck discovered off the Aegean coast in 1900, far more elaborate machines were used alongside counting boards and abacuses. Besides the spectacular art treasures recovered, a corroded metal lump remained unnoticed for a long time. It was not until the middle of the twentieth century that it became apparent that it was an 82-piece machine made of gears, axles, and other mechanical components. Through certain presettings and the turning of a hand crank, the machine, built around the year 120 BC, was able to perform the most complicated astronomical calculations.

The discovery of this machine was a stroke of luck; no one knows what other amazing machines were invented in antiquity and lost forever. But one thing is certain: from the decline of Alexandria as a centre for knowledge and throughout the Middle Ages, there was no longer any significant interest in finding mechanical ways to carry out time-consuming calculations. It was not until the seventeenth century, when the sciences took on a new lease of life, that further efforts were made to study mechanical calculating aids. There were also very practical reasons for this: the development of industry, shipping, banking, and war technology required ever more complex calculations, and the probability of calculation errors, which could have fatal consequences, increased with the number of steps in the calculation.

© The Author(s), under exclusive license to Springer Nature
Switzerland AG 2023
L. Jaeger, *Women of Genius in Science*,
https://doi.org/10.1007/978-3-031-23926-7_7

One of the scholars who worked on calculating machines was Gottfried Wilhelm Leibniz (1646–1716). The integral calculus he had invented tended to spawn endless columns of numbers:

> It is unworthy to waste the time of outstanding people with menial arithmetic work, because when a machine is used even the most simple-minded person can write down the results with certainty[1]

The graduated wheel designed by Leibniz and known as the Stepped Reckoner was the first device to handle all four basic arithmetic operations. Over the years, there were further improvements and innovations. The first modern computers based on electronic components were developed for the military in the 1940s.

There is certainly some justification for considering the introduction of electronic components as a turning point in the transition from the calculating machine to the computer. But what is easily overlooked is a fundamental change in objective. The decisive step was not so much the way the components work, but rather the idea of having a machine that could not only add, subtract, multiply, and divide numbers, but also process data of a more general nature. The collaboration between Ada Lovelace and Charles Babbage reflects precisely this leap from mechanical calculating machine to the computer.

- When he met Lovelace, Babbage (1791–1871) had already been working on his *Analytical Engine* for over ten years, with the aim of getting it to calculate endless columns of numbers for mathematical tables.
- Lovelace realised that Babbage's machine could also be used for nonnumerical tasks. She was the first person to devise and publish a specific algorithm.

* * *

Ada Augusta Lovelace grew up in a dysfunctional family. Her father was the poet known as Lord Byron, George Gordon Byron. In 1813, the 25-year-old Byron had begun a passionate affair with his half-sister Augusta. Two years later, to avoid scandal, he married the baroness Anne Isabella Milbanke, but continued the relationship with his half-sister.

---

[1] Quoted from Karl Popp and Erwin Stein (eds.), *Gottfried Wilhelm Leibniz. The Work of the Great Polymath as Philosopher, Mathematician, Physicist, Technician.* Hanover: Schlütersche (2000), p. 84.

The official Lady Byron had received an excellent education in her parental home; having a sense for logical connections, she was particularly interested in mathematics and astronomy. The marriage between the coolly rational woman and the highly emotional and adventurous poet was marked by fierce quarrels, and probably also physical clashes. On 10 December 1815, Ada was born as the couple's first and only child. Lord Byron was disappointed; he had expected a "*glorious boy*". One month after Ada's birth, the marriage finally broke down. The separation from his wife completely ruined Lord Byron's reputation and he left England. After being stationed in Switzerland and Italy, he fought against the Turks as a high-ranking commander in the Greek War of Independence and lost his life there at the age of only 36. Multiple bloodletting to treat a fever had robbed him of his last powers of resistance.

When her father died, Ada was eight years old, and the fact that she already had some burdens to bear at that time was not only due to her weak health. Even her first name was an affront: she had been named Augusta after her father's half-sister, with whom he had had an incestuous relationship. She had never consciously seen her father, and her mother prevented, with the utmost determination, any possible contact with him. While Lord Byron's poems enjoyed great popularity throughout Europe, Ada was not allowed to read his books or look at portraits of him. All her life she longed for her father and she later named her two sons Byron and Gordon. But her desire was apparently not reciprocated. Lord Byron does not seem to have given much thought to his only legitimate child. He did dedicate a few lines to her in a poem describing his journey into exile, but the strange punctuation of the first two lines makes the meaning at least questionable. If the exclamation mark and question mark were reversed, one might possibly consider it an affectionate text:

> Is thy face like thy mother's, my fair child! ADA! sole daughter of my house and heart? When last I saw thy young blue eyes they smiled, And then we parted,—not as now we part, But with a hope—

Ada's mother placed her child in her own mother's care and had other relatives and acquaintances—called "the Furies" by Ada—report regularly on her daughter's development. Ada received private lessons six days a week from six in the morning until dinner. Obsessed with preventing any contact between Ada and her father and any inclination towards poetry, her mother did her best to encourage Ada's interest in mathematics and logic. Her mathematics lessons began at the age of four, followed later by geography, astronomy, technical drawing, several languages, and music. Poetry was completely cut out of this curriculum. So Ada was caught up in a web of manipulation, yet she seized the opportunities offered to her and learned the science subjects with great interest.

The year1833 was a fateful one for Ada. As was customary for aristocratic young women, she was presented at court at the age of seventeen. Her brilliant mind made a great impression and brought her into contact with important men of her day, including Charles Dickens, who was already one of the most famous and respected writers at the time, and Michael Faraday, with whom she engaged in an extensive exchange about electromagnetism and the associated experiments. These were contacts that she would later use very well for her further education and scientific work. For example, she had frequent discussions with Faraday on the question of scientific method. But it was an encounter with a woman that was to change her life decisively.

* * *

Mary Somerville (1780–1872), whose portrait is now on the Royal Bank of Scotland's £10 note, was one of the great female scientists of the nineteenth century. She had to acquire her knowledge herself, because her family saw education as harmful for girls. Somerville became an expert on Newton, and like Émilie du Châtelet, was critical of Newton's mathematical exposition of the infinitesimal calculus. In 1826, Somerville published the results of her experiments on the relationship between light and magnetism.[2]

Instead of continuing to develop her scientific talent, she decided to increase her readership, and thus her income, by translating the works of other scholars into more generally understandable language. In 1831, her rather free translation of Pierre-Simon Laplace's five-volume work *Mécanique céleste* appeared under the title *The Mechanism of the Heavens*. This popularisation of the latest scientific results proved lucrative. Somerville's book *On the Connexion of the Physical Sciences* sold 15 000 copies. It was translated into German and Italian and was also published in the United States. In England alone, it went through ten editions and was the publisher's most successful scientific book for over a quarter of a century; it was not until 1859 that Charles Darwin's *The Origin of Species* would eventually surpass it.

Somerville continued to write general science books, corresponded with leading scholars, and took part in the current scientific debates. When she and Caroline Herschel were jointly elected as the first female honorary members of the *Royal Astronomical Society in* 1835, this was one of the highlights of her

---

[2] Mary Somerville, The Magnetic Properties of the Violet Rays of the Solar Spectrum, Proceedings of the Royal Society (1826).

life. But despite all the many other honours she received, it may be assumed that Somerville's scientific achievements fell well short of her potential, and at the end of her life, she regretted having neglected her own research.

* * *

Mary Somerville became a role model for Ada Byron, and also a reliable friend and knowledgeable conversation partner. It was Somerville who introduced Ada to Charles Babbage in June 1833. The mathematician, philosopher, and mechanical engineer was 42 years old at the time and had already been working for over ten years on a mechanical calculating machine composed of thousands of precision gears. To Babbage's great delight, Ada Byron was one of the few people who immediately grasped how his calculating machine worked. But before their collaboration could gain momentum, Ada had to navigate a few personal problems.

For one thing, she was in poor health throughout her life and regularly consumed opium, which was a common drug at the time. Indeed, she was addicted to this substance, something that was probably not unusual in her social class. However, the fact that she began an affair with one of her tutors in 1833, despite all her mother's precautions, could have had disastrous consequences for her. Shortly before the lovers were about to elope, the family took action and ended the relationship. In 1835, Ada was married to Baron William King, ten years her senior. As Lady King, she was no longer under her mother's thumb, but she had social duties to fulfil and hardly any time left for her studies and correspondence with scientists. Moreover, it was not easy for her to get hold of the latest books. As a member of the *Royal Society*, she should have had access to its extensive library, but women were only admitted there after 1945. So she asked her husband to join this institution and to copy the interesting passages for her. Despite his support, Ada King was unhappy in her marriage. At nineteen years old, she wrote in her diary:

> I believe that nothing short of close and intense study of subjects of a scientific nature can stop my imagination from going haywire and fill the void that experiential hunger has left in my mind

In the four years from 1836 to 1839, Ada King had three children. During this time, her husband was made an earl. Lovelace was chosen as the name of the new peerage, because Ada was a descendant of the Lovelace barons, whose line had died out a hundred years earlier. Thus Lady King became Lady Lovelace. After the birth of her third child, Ada Lovelace turned back to science and especially to working with Charles Babbage, with whom she had remained in contact. Indeed, Ada became the inventor's most important confidante, and in time, Babbage's project to build a calculating machine would become their joint project. He admired her as the "enchantress of numbers", and she revered him as the pioneering inventor of the first steam-powered calculating machine.

Ada Lovelace also exchanged ideas with Augustus De Morgan, a good friend of Babbage's and professor of mathematics at University College in London. In formal logic, and thus also in today's computer programming, De Morgan's two laws play a central role. In the early 1840s, De Morgan motivated Ada Lovelace to work on complex analysis and number theory. On this basis, analytical results beyond pure numerical mathematics and the four basic arithmetical operations could be described. This marked a milestone in information technology, because it meant a leap from merely processing numbers to calculating complicated formulae with the help of symbols.

$$* \quad * \quad *$$

The machine that Charles Babbage had been working on since the early 1820s was the *Difference Engine.* Its mode of operation made use of the mathematical theorem that the n th derivatives of an n th-degree function are constant. If this n th derivative of a function is known, all further values of the function can be calculated by repeated addition. Babbage's machine was supposed to handle these additions automatically in long series of loops. But in 1832, after a total of ten years of construction, only a fraction of the machine, planned with about 25 000 parts and measuring 2.6 × 2.3 × 1.0m, was actually working. The engineer John Clement had succeeded in completing a section of the machine with 2000 elements that could calculate quadratic functions. A year later, a dispute arose between him and Babbage, and in 1834 the British government pulled the plug. Over the years, it had funded the construction of the machine to the tune of a staggering £17 000, the equivalent of two warships or twenty locomotives. If Babbage had been successful, this investment would have been worthwhile: from 1792, for example, the French mathematician and hydraulic engineer Gaspard de Prony, together with about 80 mathematicians and assistants, had worked for nine years on logarithmic tables and trigonometric tables. In the end, only parts of the planned eighteen volume work were actually published, as the costs had gone off the scale.

In England, too, the hopes of producing the required numbers in a manageable time were dashed. Automation with Babbage's machine did not seem to be a solution: the construction was too complex, the mechanical parts too prone to jamming (although we know today that the planning was flawless and the machine would have worked). In 1842, the *Difference Engine* was finally laid to rest.

Babbage, who had also invested large sums of money privately in his *Difference Engine*, turned to a similar project in 1834, one year after his first encounter with Ada. It was a project he had already had in mind for some time: the *Analytical Engine*. This calculating machine was to be several metres high and driven by a steam engine. The special feature: it would be controlled by punched cards. The Frenchman Joseph-Marie Jacquard had invented punched cards in the form of wooden plates in 1805 in order to be able to automate the weaving of complicated patterns on steam-powered looms. From the information "hole or non-hole," the weaving machine was told whether or not to lift certain warp threads. This application of the 0-or-1 principle via punched cards was used as an information carrier until the 1970s.

Babbage transferred Jaquard's idea to his *Analytical Engine*. Cards made of paper were supposed to give his machine commands as to which cogs would continue to turn during a calculation and which would stand still. Ada Lovelace put it this way:

> We may say most aptly that the Analytical Engine weaves algebraical patterns just as the Jacquard loom weaves flowers and leaves[3]

He estimated that the calculating machine would take three minutes to multiply two 20-digit numbers together. For such an arithmetic operation, twenty times twenty, i.e., 400 multiplications of single-digit numbers must be carried out, and finally twenty numbers with up to forty digits must be added together. Some people, although not many, can also do this amount of arithmetic in their heads in three minutes. But a machine would be able to do it without pause and without error.

* * *

Michael Fothe, professor of the didactics of computer science and mathematics at the University of Jena, has described in detail how Charles Babbage's (and Ada Lovelace's) *Analytical Engine* was designed.[4] Here are some of its features:

---

[3] Frederico Menabrea, Ada Lovelace, *Grundriß der von Charles Babbage erfundenen Analytical Engine*, in Bernhard Dotzler (ed.), *Babbage's Rechen-Automate—Ausgewählte Schriften*, Computerkultur Band VI, Springer Wien/New York (1996), p. 335.

[4] Michael Fothe, *Computer Science Has History!* in: Heinrich C. Mayr, Martin Pinzger (eds.), *Informatics 2016, Lecture Notes in Informatics (LNI)*, Gesellschaft für Informatik, Bonn (2016) pp. 1.909–1.915.

- The machine processes numbers with 200 digits and more.
- There is a kind of "warehouse" where punched cards with intermediate results are stored, and a "mill" where the actual calculation steps are carried out.
- The information is carried via operational cards, variable cards, and reception cards.
- The operational cards determine which of the basic arithmetic operations are to be applied. They remain active as long as the same arithmetic step is being repeated.
- The variable cards contain information about the variables that the operator feeds to the machine.
- The reception cards are also variable cards, receiving their values from the calculator.
- A single arithmetic operation requires two cards, an operational card and a reception card. The variable cards contain the numbers that are to be connected via an arithmetic operation.
- A return system enables the programming of loops and branches.

Even though Babbage never published his own detailed description of his machine, his work was followed with great interest in Europe. At a scientific congress in Turin in 1840, he explained the concept of his *Analytical Engine*, and a member of the audience, the mathematician, engineer, and later Prime Minister of Italy Federico Menabrea (1809–1896), published a twenty page article on the subject as presented by Babbage in 1842.[5] So that the article, written in French, could also be disseminated in the English-speaking world, Lovelace translated it and, at Babbage's suggestion, included extensive additions under the pseudonym A.A.L. Ultimately, her seven explanatory notes made the article three times longer than Menabrea's original text and it became the most comprehensive contemporary account of Babbage's idea in English.

Today, there is no way to find out how far Lovelace was involved in the development of the *Analytical Engine*. Were the punch cards her idea? How fully did she realise the power of algorithms? Or was she just a gifted self-promoter, good at making an impression? In any case, her additions to Menabrea's publication impressively demonstrate her independent thinking. While Babbage saw the usefulness of his machine more in multiplying numbers without error, Ada Lovelace looked beyond the purely numerical possibilities of this machine. She explained how Babbage's analytical

---

[5] Luigi Frederico Menabrea, *Notions sur la Machine Analytique de M. Charles Babbage*, Bibliothèque universelle de Genève, nouvelle série 41 (1842) pp. 352–76.

machine, if constructed, could also perform sequential operations on letters, musical notes, and pictures. With this vision, she anticipated the concept of today's computer science a hundred years before Konrad Zuse built the first computer. With regard to music, she wrote:

> Supposing, for instance, that the fundamental relations of pitched sounds in the science of harmony and of musical composition were susceptible of such expression and adaptations, the Engine might compose elaborate and scientific pieces of music of any degree of complexity or extent[6]

Lovelace had recognised that the concept of the *analytical engine* would be able to process basically any information or solve the associated problem, as long as the information and the required algorithms could be translated into mathematical notation. The English computer scientist Alan Turing proved in the 1930s that this was indeed the case for musical notes and languages. This kind of translation inspired by Lovelace has been called computer science since the twentieth century. There are two components at the centre of this discipline:

- the transformation of complex information such as language or music into mathematics,
- the use of algorithms, i.e., finite series of unambiguous instructions for processing the input information.

Lovelace also provided an example of such an algorithm in her supplements. She proved that the *Analytical Engine* could calculate the Bernoulli number sequence (whose existence she had learnt from De Morgan). A diagram illustrated the algorithm and the corresponding table extended the necessary calculational steps to a mathematical principle in which decimal numbers were converted into binary numbers (finite sequences of 0's and 1's) and elementary commands were compiled in a list. Lovelace's design for the automated calculation of Bernoulli numbers is now regarded as the first computer program in history and she herself as the inventor of this software. In her additions to Menabrea's article, she wrote:

> New language has emerged alongside rich scientific inquiry and research

---

[6] Ada Lovelace, *Notes to a Sketch of the Analytical Engine Invented by Charles Babbage, by L.F. Menabrea*, Scientific Memoirs, 3, London (1843).

But Lovelace also showed the limits of the machine: it would never be able to produce something creative on its own. This machine could only ever do what people had asked it to do.

Measured against the state of knowledge at the time, Lovelace's ideas and arguments were visionary and one of the most exciting intellectual achievements of the nineteenth century. However, although the idea of the modern computer received some recognition among experts, it was not pursued. There were two main reasons for this:

- After the demise of his *Difference Engine,* which was generally regarded as a failure, Babbage was unable to find financial backers who would have made it possible to complete the *Analytical Engine.* Only the concept of this machine exists; it was never built.
- Lovelace died of cancer on 27 November 1852 at the age of only 36.

Ada Lovelace was buried at her request next to her father, who had also died at the age of 36, in the church of St Mary Magdalene in Hucknall, Nottinghamshire.

* * *

One can only speculate how far computer technology might have come by today if Lovelace had lived a few decades longer. The fact that her notes were quickly forgotten after her death could be due to her penchant for metaphysics. She was, for example, a follower of the pseudosciences of phrenology, which claims to deduce mental capacities by measuring the skull, and mesmerism, according to which there is a kind of animal magnetism possessed by all living beings. In the scientific community of the day, which already had a hard time taking any woman's thoughts seriously anyway, this was probably one reason why the significance of Lovelace's elaboration on the possibilities of calculating machines was dismissed as nonsense. But it was precisely in the mixture of boundless imagination and strictly rational thinking that she had inherited from both her parents that lay the key to her extraordinary perspective on Babbage's *Analytical Engine.* That Ada Lovelace was at home in both worlds is reflected in a frequently quoted (but unsourced) remark she once made:

> Those who have learned to walk on the threshold of the unknown worlds, by means of what are commonly termed par excellence the exact sciences, may then, with the fair white wings of Imagination hope to soar further into the unexplored amidst which we live

Ada Lovelace was certainly not an easy person to live with. A certain manic behaviour has come to light in the historical research about her over the past two centuries. She tried her husband's patience with several affairs and an extraordinarily lavish lifestyle, even for her those moving in her circles. She tried to compensate for her constant lack of money by betting on horses, but this only increased her debts. One reason for her interest in Babbage's machine may have been the hope that an automated system could improve her winnings. She also suffered from an overconfidence that would be fatal for any gambler:

> I was born into the world as a prophetess and this conviction fills me with humility, trembling and quaking[7]

In the end, she was abandoned by those with whom she spent most of her time. Charles Babbage refused any further collaboration with her when she tried to reverse their relationship and effectively make him *her* assistant, even though he was the one who had worked tirelessly for decades designing calculating machines. And shortly before her death, when she was already bedridden, her husband also broke off contact with her. She had confessed something to him—it is not known what "sin" she committed—but it made him leave the room immediately and never return to her.

* * *

One hundred years after Lovelace's death, the basic principle of the computer was developed a second time. The concept had remained the same, only the technical possibilities had now progressed considerably and the scientific community had become more open-minded. The political conditions for building expensive computers were also favourable: the first major application of computers was to calculate processes occurring in atomic and hydrogen bombs.

In 1953, Lovelace's notes were republished for the first time in a book on digital data processing.[8] This triggered our modern perception of her as one of the masterminds of what we now call computer science. Today she is often referred to as the "prophet of the computer age". Consequently, an important computer language ADA, designed by the French computer scientist Jean Ichbiah, was named after her in the 1970s. Her dazzling personality

---

[7] Ada Lovelace, November 1844; here, after Agnes Imhof, Die geniale Rebellin: Ada Lovelace—Sie stürzte sich ins Leben und revolutionierte die Mathematik, Piper Taschenbuch (2022).

[8] Bertram V. Bowden (ed.), *Faster than Thought—A Symposium on Digital Computing Machines*, Pitman (1953).

makes it easy to project the most diverse views onto her. But regardless of how her achievement is assessed, she was undeniably a child of her time: fascinated by the possibilities of the technological age, mechanisation, and steam engines, and in addition, exceptionally creative and full of curiosity regarding the promises of the modern world.

8

Sofia Kovalevskaya (1850–1891): An Irrepressible Russian Mathematician

Until well into the twentieth century, women had to fight for their right to education. The few who managed to achieve something significant in their subject of interest either educated themselves with great effort or enjoyed the rare good fortune of being born to financially well off and at the same time progressive parents who also allowed their daughters private lessons.

In Sofia Kovalevskaya's case, all these things came together. Her parents provided her with a good education through private tutors, even though they were hired primarily for her younger brother. But there is also a story about how Sofia, when she was about eleven years old, would stare at the walls of her room for hours trying to make sense of the numbers and formulae visible on them—for when her parents' manor was being renovated, they hadn't ordered enough wallpaper and had unceremoniously pasted her father's lecture notes on differential and integral calculus on the walls in the least important room of the house. Kovalevskaya later wrote:

> I must confess that at that time I understood virtually nothing about it, but it was as if an irresistible force drew me to this occupation. As a result of my persistent contemplation, I learned many passages by heart, and some formulas became so firmly engraved in my memory that they left deep traces in it.[1]

This can be said to characterise Sofia Kovalevskaya's entire life. She was a gifted mathematician and at the same time an entrepreneur, a writer, an

---

[1] Sofja Kowalewskaja, Autobiographical Sketch, *Deutsche Rundschau*, 108 (1901), pp. 152–160.

enthusiastic nihilist, a passionate lover, and a committed advocate of women's rights. She was the child of wealthy parents, but was herself bankrupt and suffered much of her life in precarious financial circumstances. Reserved and tormented by self-doubt, she was nevertheless a gifted networker. At a time when women's freedom of movement was generally limited to the very minimum, she travelled back and forth—alone!—across Europe for years. No woman before her had succeeded in gaining a doctorate in mathematics in the utterly misogynistic university structures of the day, and subsequently obtaining a professorship. What she experienced, achieved, and suffered would be enough to fill several lifetimes.

* * *

Sofia Kovalevskaya was born Sofia Korvin-Krukovsky in Moscow on 15 January 1850. Her parents had fervently hoped for a son six years after the birth of their first child Anna, but their desire for a third child and heir would not come true for them for another five years. Little Sofia often heard her nanny say that her parents had prepared everything for the birth of a boy and that the fact that she was a girl had been a great disappointment. Writing about her youth in her memoirs, published in 1897, she writes:

> Thanks to these conversations, I became convinced at an early stage that I was not one of the favourites, and this had an effect on my character; I developed more and more shyness and reticence.[2]

Sofia's father was of Belarusian and Polish descent, a lieutenant general in the Russian army, and belonged to the lower nobility. Sofia grew up on his country estate in Palibino in the north of present-day Belarus. Sofia's mother Elisabeth was the daughter of the quite well-known German topographer and surveyor Friedrich Schubert and granddaughter of the St. Petersburg astronomer Theodor von Schubert.

Private tutors were hired primarily for Feodor, the baby of the family, but Sofia was allowed to attend the classes. When she discovered her love of mathematics, her father felt that his daughter's thirst for knowledge exceeded the bounds of propriety and forbade her to attend further classes. Sofia Kovalevskaya's mathematical career could have ended here, before it had really begun.

By chance, one of the family's neighbours was the physics professor Nikolai Tyrtov. When he gave Sofia's father his recently published "Textbook of

---

[2] Sofja Kowalewskaja, *Jugenderinnerungen*, Translated into German by Louise Flachs-Fokschaneanu. Available at: Project-Gutenberg.org.

Elementary Physics", Sofia tried to understand its contents, but failed at first because she didn't have a sufficient knowledge of the trigonometric functions used in optics. She thus derived the formulae on her own and told Tyrtov at the next opportunity that she had read his book with great interest. At first, Tyrtov thought this must be "vain boasting", but he soon came to recognise Sofia's unusual talent. He convinced her father to allow his daughter further lessons. Kovalevskaya later reported in her autobiographical writings that she was taught calculus and other subjects in St Petersburg, where her family usually spent the season. Sofia Kovalevskaya never attended a grammar school, but thanks to intensive self-study and private lessons, she had already left her peers far behind in the field of mathematics as a teenager.

But Sofia was also a very normal young girl. During one of her stays in St. Petersburg, the family came into contact with the writer Fyodor Dostoevsky. Under the watchful eye of her parents, a complicated web of relationships developed. The very young Sofia fell in love with the famous writer, showing him her own poems and playing the piano for him. Dostoevsky seemed unable to decide between the sisters for some time but finally proposed to Anna. However, Anna refused—Dostoevsky's political views were not progressive enough for her.

* * *

Sofia and her sister Anna, eight years her senior, with whom she had little in common as a child, became closest confidantes. Anna introduced Sofia to the ideas of Russian nihilism. The name of this school of thought is misleading, for its adherents did not believe "in nothing" but, on the contrary, enthusiastically believed that their country should be purposefully and fundamentally renewed. Nihilism was a response to the untenable social conditions in the Tsarist empire and rejected all authority and roles perceived as arbitrary. One of its most important goals was to strengthen women's rights and provide them with educational opportunities.

The ideas put forward by the nihilist movement offered Sofia and Anna a way to get a good deal closer to their desire for a self-determined life. Sofia's dream was to study abroad, because universities in Russia, as in almost every country in the world, were closed to women. But even a trip abroad required her father's consent. Since a woman was transferred from her father's care to that of her husband when she married, marriages of convenience were often used within the nihilist movement. The purpose of such a union was to give a woman *carte blanche* to make her own decisions and especially to travel abroad.

The 18-year-old Sofia Korwin-Krukowski also uses this way out to escape from her father's authority. In 1868, she entered into a sham marriage with Vladimir Kovalevsky, a lawyer, palaeontology student, and radical nihilist, eight years her senior. Sofia was now called Sofia Kovalevskaya and could pursue her own plans. Sofia and Vladimir became good friends, and in their letters they addressed each other as "brother" and "sister." However, living together as a married couple was out of the question for both of them, at least at first.

The Kovalevskys left Russia together in 1869. Many doors opened for Vladimir in Western Europe. He studied the fossil collections at German universities and worked on a family tree of odd-toed ungulates at a time when Charles Darwin's theories were still highly controversial. As a luminary in this field of research, he met Darwin personally and was later the first to translate his works into Russian and publish them.

Sofia's fight proved more challenging. Her goal had been to find a little more freedom in Western Europe with Vladimir and her sister Anna. Initially, she hoped to train as a doctor at Zurich University, which in 1867 was the first university to officially admit women to study, and then go on to Siberia to care for exiles there. Their first stop was Vienna, although they soon had to leave for financial reasons. Next, the trio travelled to Heidelberg, where Sofia returned to the idea of studying mathematics. But what was already possible in Zurich—women enrolling officially and with full rights—was still considered a downright ridiculous idea at all other European universities.

* * *

In Heidelberg, it was up to individual professors to decide whether or not to tolerate women in their lectures, without, of course, any entitlement to further supervision or even an examination arising from such permission. However, Kovalevskaya succeeded in convincing some of the most important scientists of the day to allow her into their lecture halls: the mathematicians Leo Koenigsberger and Emil du Bois-Reymond, the physicists Hermann Helmholtz and Gustav Robert Kirchhoff, and the chemist Robert Bunsen. The lectures with Helmholtz and Kirchhoff awakened in her a fascination for theoretical mechanics, and during the break in the semester, she continued to work intensively in this field. To further Kovalevskaya's interest in this field of research, Koenigsberger advised her to join his teacher Karl Weierstrass in Berlin. In the autumn of 1870, the three of them parted company; Anna went to Paris, where she fought as a revolutionary on the Paris barricades, while Sofia and Vladimir Kovalevsky moved to the Prussian capital. But the situation in Berlin were even more restrictive than in Heidelberg: women were strictly forbidden to enter universities in Berlin.

Kovalevskaya made a personal visit to Weierstrass in his flat, but the great mathematician did not even take note of her letter of recommendation from Heidelberg. Instead, he gave her a few exercises for advanced students to take home. When he received the solutions from Kovalevskaya a week later, he immediately recognised the young woman's extraordinary talent. But even the enthusiastic advocacy of this world-famous mathematician could not convince his colleagues to allow Sofia Kovalevskaya to enter the Berlin University; there wasn't even any discussion of official enrolment (Prussia was one of the last German states to allow women to enrol, in 1908).

Weierstrass saw no other way to support Sofia Kovalevskaya than to offer her private lessons. For four years, she studied with the internationally renowned mathematician, helping him with his own mathematical studies as her skills progressed and eventually becoming his confidante. The relationship between Kovalevskaya and Weierstrass, a bachelor 35 years her senior, was one of the greatest strokes of luck in the history of mathematics. Weierstrass enjoyed the exchange of ideas with the brilliant mathematician, and Kovalevskaya received the best possible education from her mentor. Contrary to all known rules of etiquette at the time, the two were even on familiar terms with each other. On 17 June 1875, when Kovalevskaya left Berlin, Weierstrass wrote to her:

> I have never found anyone who has shown me such an understanding of the highest aims of science and such a joyful acceptance of my views and principles as you have![3]

Even though Weierstrass' letters strike romantic notes in places, there was never a love affair between Kovalevskaya and Weierstrass. Many years later, Kovalevskaya wrote about her teacher:

> He had the most important influence on my entire mathematical career. He gave me the definitive, unchanging direction that I followed in my scientific work from then on, and all my work is written in the spirit of Weierstrass' ideas.[4]

* * *

---

[3] Reinhard Bölling (ed.), *Briefwechsel zwischen Karl Weierstrass und Sofja Kowalewskaja*, Akademie Verlag Berlin (1993).

[4] Sofja Kowalewskaja, Autobiographical Sketch, *Deutsche Rundschau*, 108 (1901), pp. 152–160.

Weierstrass had great difficulty finding a university for his student that would grant her a doctorate. His colleagues in Berlin had already declined. Professors in other cities also feared a precedent and refused. Weierstrass's referral to Gauss, who had noted with regret in 1837 that it had been an oversight not to award the mathematician Sophie Germain a doctorate during her lifetime, also failed to bear fruit.

Only the University of Göttingen turned out to be more open-minded. In the summer of 1874, Sofia Kovalevskaya submitted three papers there. Weierstrass was not the only one who thought that each of them would have sufficed for an excellent doctorate. The Göttingen referee Ernst Schering also noted this.

- "Zur Theorie der partiellen Differentialgleichungen" (On the Theory of Partial Differential Equations) was the first of her contributions and was seen as the core text of her dissertation. Leonhard Euler (1707–1783) had been the first to discover this type of differential equation and solve it in a special case. After Joseph-Louis Lagrange (1736–1813) had described another case, certain properties of a third type had remained open. It was only known that there would be no further cases within calculus— Kovalevskaya succeeded in solving this third case and thus brought research in this part of mathematics to a close. Today, research is again being done on partial differential equations, but they are a new type that can no longer be represented by a Taylor expansion and therefore do not belong to the class of analytical equations. That Kovalevskaya's work was printed in the Journal of Pure and Applied Mathematics was an extraordinary honour.[5]
- Her second work dealt with the reduction of certain Abelian integrals to simpler elliptic integrals.[6] The relationship between Abelian and elliptic integrals was a major focus of nineteenth century analysis; her mentor Weierstrass had also spent much time on this topic.
- Kovalevskaya's third contribution dealt with a topic in astronomy. In it, she further developed work on the shape of Saturn's rings, first studied by Laplace. It is not published until more than a decade later.[7]

The first paper alone catapulted Kovalevskaya straight to the top at the age of just 24. Now the professors in Göttingen were in great trouble,

---

[5] Sofja Kowalevskaja, On the Theory of Partial Differential Equations, *Journal of Pure and Applied Mathematics*, 80 (1875), pp. 1–32.

[6] Sofja Kowalewskaja, Ueber die Reduktion gewisser Abel'scher Functionen auf elliptische Functionen, *Acta Mathematica*, 4 (1884), pp. 393–414.

[7] Sofja Kowalevskaja, Additions and Remarks Regarding Laplace's Investigation of the Shape of Saturn's Rings, *Astronomische Nachrichten*, 111 (1885), pp. 37–48.

because they still had to put her through an oral examination to be able to award the doctorate. The procedure of presenting herself personally to all the professors—most of whom were reactionary—and taking an oral examination in front of the assembled team could have led to a scandal. Moreover, Kovalevskaya still only spoke broken German, and Weierstrass feared that the shy young woman would not be up to an examination.[8] It was eventually agreed that she would receive her doctorate in August 1874 without a personal appearance, i.e., *in absentia*. So, even though the name Sofia Kovalevskaya is inseparably linked with Göttingen University, she never set foot in the city.

If Kovalevskaya had been a man, the mathematics faculties of Europe would immediately have offered her a professorship and tried to outbid each other to give her a post. But it turned out that, for a woman, a more advanced mathematical career was inconceivable, either in Germany or in any other country—her doctorate alone was unique in the history of mathematics. Had Kovalevskaya really believed that a university would offer her a position? The disappointment regarding her lack of a future in Western Europe was a major psychological blow.

* * *

Sofia Kovalevskaya returned to Russia in 1874, and Vladimir was also back in his homeland. He had spent most of Sofia's Berlin years travelling through Europe, studying palaeontology and earning his doctorate in Jena. But his hopes of a professorship had also been dashed, like those of his wife; in his case, it was his radical convictions that had closed doors. Not even Kovalevskaya's fallback plan of teaching in Russia could be carried out. She lacked a Russian mathematics diploma and her world-class Göttingen doctorate was not recognised as a substitute. Since as a woman she had no access to Russian universities, she could not even take the required exam either—which she would certainly have been able to pass with the greatest ease. So, the couple were stranded and had to find new sources of income. For six years, mathematics would take a back seat for Sofia.

They kept their heads above water with various projects. Kovalevskaya tried her hand as a writer and theatre reviewer, with some success; and in an overheated real estate market, the couple's investment activities seemed to offer further prospects for success.

---

[8] Weierstrass' letter to the Göttingen professor and former student Lazarus Fuchs of 3 July 1874 is available on the internet at histmath-heidelberg.de/txt/Kowalewsky/promotion.htm.

During my entire stay in Russia, I did not write a single independent work. The only thing that stimulated me scientifically to some extent was the correspondence and exchange of ideas with my dear teacher Weierstrass.[9]

During this time Sofia and Vladimir Kovalevsky began to live together as a couple. Perhaps the death of Kovalevskaya's father contributed to this change in their relationship. Shortly after the birth of their daughter Sofia in 1878, the speculative bubble burst and Sofia and Vladimir went bankrupt.

In this crisis, Kovalevskaya turned to mathematics once again and quickly fell back under its spell. At the beginning of 1880, she presented her still unpublished dissertation on the classes of Abelian integrals at the Congress of Naturalists and Physicians in St. Petersburg. Although six years had passed, the topic was still red hot and Kovalevskaya received a lot of attention. On this occasion she met the Swede Magnus Mittag–Leffler. Like her, he was a student of Weierstrass, and he was the one who would give her life a significant boost a few years later.

Sofia Kovalevskaya had already travelled a lot during her short life. Among other things, she had met the English writer George Eliot in London in 1869, and had cared for the injured revolutionaries, including women, alongside her sister Anna during the uprising of the Paris Commune in the spring of 1871. After the six years in Russia, she began once again rushing restlessly through France, Italy, and above all Germany to renew old acquaintances and connect with current research.

- First, she moved to Moscow with her husband and child, where she attended the events organised by the Moscow Mathematical Society.
- In the spring of 1881, she finally separated from her husband and travelled to Berlin alone with her daughter.
- She entrusted her daughter to a good friend whom she had met during her studies. Her friend and daughter returned to Russia, while Sofia Kovalevskaya remained in Berlin.
- From the end of 1881, she was in Paris. In May 1882, Gösta Mittag–Leffler visited Kovalevskaya in Paris and established contact with the most important French mathematicians. Just two months later, she was elected to the Paris Mathematical Society.

While Kovalevskaya was expanding her network, her husband, who had stayed behind in Russia, became more and more emotionally unstable. Speculation in the oil business caused him and several friends and relatives to

---

[9] Sofja Kowalewskaja, *Autobiographical Sketch*, Deutsche Rundschau, 108 (1901), pp. 152–160.

lose large sums of money and, suspected of embezzling company funds, he committed suicide in April 1883. Sofia was deeply shocked by the news, but the very fact that she was now a widow would soon lead to a new chapter in her life.

* * *

In 1878, Stockholm University was founded and Gösta Mittag–Leffler was appointed its first mathematics professor. In contrast to time-honoured universities like Uppsala, which were also steeped in tradition, the new university was intended to bring a breath of fresh air to Sweden's educational landscape. However, this desire for change and renewal did not mean that women would be granted the same rights as men.

Mittag–Leffler was only with difficulty able to get Kovalevskaya a position in the mathematics faculty, and even then only because she was granted some freedoms as a widow according to the prevailing social customs. As a wife or living separately from her husband, it would have been quite impossible for her to accept a position. Kovalevskaya was already quite famous and her arrival in Sweden at the end of 1883 was a sensation—all the newspapers in Sweden reported on it. However, the fact that she was perceived more as a kind of freak than as a champion of women's rights is well illustrated by the assessment of the contemporary commentator August Strindberg in a newspaper article from 1884:

> A female mathematics professor is a dangerous and unpleasant phenomenon; it is safe to say, something of a monstrosity.

The Stockholm professorate also received Kovalevskaya with massive resistance. At first, she had to work without a salary, and it was only a year later that Mittag–Leffler was able to secure her appointment as an associate professor for a limited period of five years. Today, her place in the university hierarchy would be that of an *assistant professor*. Thanks to this position, Sofia Kovalevskaya enjoyed financial security for the first time in many years. From now on, things were looking up for her professionally. She had a good network in Europe and knew how to make a name for herself. The fact that Mittag–Leffler made her an editor of the specialist journal he had founded, *Acta Mathematica*, adds to her renown.

Kovalevskaya continued her active and creative role in mathematics, while at the same time managing to remain active as a writer. She wrote a play with Gösta Mittag–Leffler's sister, the actress, novelist, and playwright Anne Charlotte Edgren-Leffler, with whom she had a close friendship.

* * *

Since her student days, Kovalevskaya had repeatedly turned her attention to the so-called rotation problem, one of the most famous questions in mathematics:

> The greatest minds have struggled with it, among them Euler, Lagrange, and Poisson. (...) In the history of mathematics we encounter few problems whose solution would be so ardently desired and on which so much effort and strength have been expended without leading to essential results.[10]

As a private lecturer, she had much free time and was able to work intensively on this problem, and in 1888 she achieved the decisive breakthrough by finding the solution for the complex motion of a spinning top. The Paris Academy of Sciences, with whom she was in close contact, encouraged her to work out her results and submit them in time to compete for the Prix Bordin. In Kovalevskaya's day, this was the most important international prize for mathematicians. The deadline was 1 June, but Kovalevskaya missed the date—for personal reasons.

Sofia Kovalevskaya was not only a political and mathematical revolutionary, she also went her own way when it came to personal relationships. Influenced by the spirit of optimism of the Russian nihilists, she insisted on self-determination in this area of her life as well. At the end of 1887, for example, she met Alfred Nobel, seventeen years her senior. The rich entrepreneur courted her, but she rejected him. There is a legend that there was no Nobel Prize for mathematics because Kovalevskaya dropped Alfred Nobel for the mathematician Mittag–Leffler. Indeed, Nobel and Mittag–Leffler did not have too good a personal relationship, but the rumour that Sofia Kovalevskaya was emotionally involved with Mittag–Leffler has no basis.

In fact, she began an affair with the polar explorer Fridtjof Nansen and, even before she learnt that he was already engaged, a new man entered her life: the Russian sociologist Maxim Kovalewski, a cousin of her late husband. She spent weeks with him in Stockholm and London in the first half of 1888. At this time, Paris was determined to award the Bordin Prize to Kovalevskaya and opened a back door for her so that she could submit her work later. However, her elaboration of the gyroscope problem was a long time coming. It was only during a summer holiday that Weierstrass spent with her and his two sisters that Kovalevskaya finally managed to finish the work.

---

[10] Sofja Kowalewskaja, Autobiographical Sketch, *Deutsche Rundschau*, 108 (1901), pp. 152–160.

In her account, which was later published under the title "On a special case of the problem of the rotation of a heavy body about a fixed point,"[11] Kovalevskaya described her discovery of the mathematical figure now known as the Kovalevskaya gyroscope, and showed that, apart from the cases considered by Euler and Lagrange, this was the only other case of the rotational motion of a rigid body whose solution functions were unambiguous and completely integrable and could therefore be solved exactly using calculus. The experts were thrilled because Kovalevskaya had solved one of the biggest problems in mathematics that remained open at the time.

Officially, the entries for the Prix Bordin are sent to Paris anonymously, but it was obvious which of the fifteen entries was Kovalevskaya's. The Parisian committee awarded her the prize and even increased its value from 3000 to 5000 francs.

$$* \quad * \quad *$$

The award of the Prix Bordin virtually forced the Stockholm professors to finally offer Kovalevskaya a full professorship and thus a lifelong income. In 1889, Kovalevskaya won the prize of the Swedish Academy and was also appointed a corresponding member of the Russian Academy of Sciences, something that was only possible in the conservative Tsarist Russia after the university statutes were changed. Her great wish, a professorship in Russia, was never fulfilled, however. She knew well the advantages of Sweden as a country, but she was homesick for Russia. Her periods of depression grew, and the many honours she received did nothing to change that. She had already written to Weierstrass at the beginning of her professorship:

I gave my first lecture today. I don't know if it was good or bad, but it was very sad to come home and feel so lonely in the wide world. In moments like that, you can feel it particularly strongly. Another stage of life behind me.[12]

When her beloved sister died in 1887 after a serious illness, she tried to overcome her pain by writing, something she had never completely given up. In 1889, her greatest literary success appeared in Stockholm: her "Jugenderinnerungen"[13] was translated into several languages. One year later she wrote:

---

[11] Sofja Kowalewskaja, *Sur le problème de la rotation d'un corps solide autour d'un point fixe,* in: Acta Mathematica, 12 (1889), pp. 177–232.

[12] Reinhard Bölling, For the First Time—A Look at a Letter from Kovalevskaya to Weierstrass, *Historia Mathematica,* 20 (1993), pp. 126–150.

[13] Kovalevskaya's memoirs of her youth were published in 1890 in the Russian Journal Westnik Jewropy. The text is available at: https://www.projekt-gutenberg.org/kowalews/jugender/chap_001.html.

All my life I have not been able to decide whether I prefer writing or mathematics. As soon as my head is tired of purely abstract considerations, I immediately feel impelled to make observations about life and to write stories. And it can be the other way round, too, that everything in life seems insignificant and indifferent to me and I am attracted only by the eternal laws of science. It is possible that I could have achieved more in one or the other of the two fields if I had devoted myself entirely to the same. But it was not possible for me to give up either of them altogether.[14]

* * *

Sofia Kovalevskaya spent her life torn between many different needs and desires. There was her longing for a peaceful private life with her daughter, but also her ambition to achieve major mathematical results. Equally important to her, however, were her literary activities and her fight for equal rights for men and women. Attending to all of these at the same time consumed her mental and physical strength. The many journeys between Russia, Stockholm, and Paris also put a strain on her health.

She had just begun a new and intensive creative phase in the field of mathematics when she caught a cold on her return from a holiday with Maxim in Nice and Cannes. The couple travelled to Sweden via Paris and Berlin, where she met with the most famous mathematicians, as she so often did. However, the cold turned into pneumonia. On 10 February 1891, she died completely unexpectedly in Stockholm at the age of only 41.

Her outstanding achievements in analysis, function theory, partial differential equations, and theoretical physics, were truly groundbreaking steps for nineteenth century mathematics. And the decisive impetus given by a number of important Russian mathematicians in the first half of the twentieth century can ultimately be traced back to her work.

Today, the Alexander von Humboldt Foundation's Sofia Kovalevskaya Award gives young foreign researchers the opportunity to set up a research group at any German research institution and carry out a research project for five years. With up to 1.65 million euros in award money, it is one of the most richly endowed science awards in Germany.

---

[14] Reinhard Bölling, *Say What You Know, Do What You Must, and Let It Be Done. Sofia Kovalewskaya—Periods of Her Life*; https://www.math.uni-potsdam.de/fileadmin/user_upload/Institutskolloquium/Boelling/Boelling_Kowalewskaja_alt.pdf.

9

Marie Curie (1867–1934): Pioneer of Nuclear Physics

Few people are aware that Marie Curie, probably the world's most famous scientist to this day, was officially Russian and thus a compatriot of Sofia Kovalevskaya. She was born Maria Salomea Skłodowska in 1867 in Warsaw, i.e., in the Russian part of the former Kingdom of Poland, which was then divided between Prussia and Russia. Her family belonged to the Polish landed gentry and had lost landed property and assets under the Russian occupation; the five Skłodowska siblings—Maria was the youngest—thus grew up in modest circumstances. However, the family remained proud of their Polish origins.

Maria was eleven years old when one of her sisters died of typhoid fever, and her mother, who had worked as a school headteacher, lost her battle against tuberculosis. She was placed in a boarding school and later attended a high school for girls, graduating at the top of her class in 1883. But the burdens of the past caught up with her and she suffered what was probably a nervous breakdown—she had periods of melancholy throughout her life. It took more than a year before Maria was able to attend the illegal "Flying University" together with her sister Bronisława, called Bronya, who was two years older. This institution, which was only legalised in 1905, gave women access to education in Warsaw for three decades.

Maria and Bronya subsequently made an agreement. Bronya went to Paris to study medicine—women had been able to enrol there since 1863—and was supported financially by Maria. In return, Bronya was to enable her sister to study two years later in a similar manner, by sending her money". Maria

© The Author(s), under exclusive license to Springer Nature Switzerland AG 2023
L. Jaeger, *Women of Genius in Science*,
https://doi.org/10.1007/978-3-031-23926-7_9

fulfilled her part of the bargain, working as a governess in various families and sending money to her sister. During this time, she wrote in a letter to a cousin:

> I would not wish such a life in hell on my worst enemy! (...) people posture on liberalism, but in reality the darkest stupidity prevails (...) My knowledge of the human species has expanded greatly here. I have learned that (...) one must have nothing to do with people whose moral standards have been lowered by wealth.[1]

It was not until the beginning of 1890 that Bronya completed her studies, and even then her financial circumstances were not sufficient to support Maria as agreed. Maria continued to work in Poland to raise the necessary funds for her own studies. During this time, she took every opportunity to further her education. This already difficult time was made even more unbearable by a budding but impossible relationship. Maria and the son of one of her employers (who later became a well-known Polish mathematician) fell in love with each other, but his parents put a stop to the relationship.

* * *

At the end of 1891, her time has finally come: at the age of 24, Maria went to Paris. Marie, as she was now called in France, found a cheap attic room in the immediate vicinity of the Sorbonne. Despite all her efforts, her education in Poland had been so poor that the physics courses at the university overwhelmed her. In addition, she had difficulties with the French language. Her life became a story of hardship; in summer her attic room was scorching hot, and in winter she had to wear all the clothes she owned, one on top of the other, to keep herself reasonably warm. Moreover, in addition to her efforts to keep up with her studies, she had to give private lessons in the evenings to make ends meet financially. It was only through perseverance and hard work that she overcame all these obstacles.

But Marie Skłodowska's diligence would pay off. She won a scholarship that allowed her to study mathematics in addition to physics. In the summer of 1893, she graduated with the highest grade in physics. Even before she came second in mathematics the following year, she received her first paid research assignment: she was to investigate the magnetic properties of various types of steel. In search of a laboratory in which she could carry out this work, Marie Skłodowska met Pierre Curie in the spring of 1894. The shy laboratory

---

[1] Fritz Vögtle and Peter Ksoll, *Marie Curie*. Rowohlt (2018), p. 14.

director at the Municipal School of Industrial Physics and Chemistry would probably go down as a nerd today. Despite his scientific and technical talent and many research successes, he was working at a second-rate institution. He wanted to support the 27-year-old graduate, but his lab was far too small to accommodate her. A tiny room was thus found in which Marie Skłodowska could carry out her experiments.

Marie Skłodowska and Pierre Curie soon shared more than a fascination for science and a belief in social responsibility. The year they met, Pierre wrote to Marie, who was eight years his junior:

> It would be a beautiful thing, nevertheless, which I hardly dare believe, to spend our lives side by side, hypnotised by our dreams: your patriotic dream, our humanitarian dream, and our scientific dream.[2]

Marie Skłodowska rejected the marriage proposal Pierre Curie made to her after only a few months, because she loved her homeland and hoped to return there. Curie was willing to follow her to Poland, but he never had to keep this promise. For when Marie went to Poland immediately after passing her mathematics exams in the summer of 1894, she discovered that, as a woman, she would not be able to work as a scientist there.

Curie convinced Skłodowska to do her doctorate in Paris. In return, the young scientist motivated her friend to set his own dissertation down on paper. For Pierre Curie, although he had already been doing research for fifteen years and had gained a very good reputation, had never taken the time to do his doctorate. So they worked on their dissertations at the same time. Once again, there was an agreement between two people, and once again, it would become apparent that the cards were unequally distributed.

In July 1895, Marie Skłodowska and Pierre Curie married. They both refused a religious ceremony, and instead of a wedding dress, Marie wore a dark blue dress that would serve as work clothes for many years to come.

* * *

The year when the Curies were married happened to coincide with the beginning of one of the most exciting chapters in the history of science. In December 1895, the German physicist Wilhelm Röntgen discovered X-rays during his laboratory experiments. The experts were astonished: these rays, caused by strong electrical discharges, were able to penetrate certain materials

---

[2] Susan Quinn, *Marie Curie: A Life*. Addison-Wesley (1996).

as though they put up no resistance, and they could also blacken photographic plates. With this discovery, an old dream of doctors came true: to be able to non-invasively visualise the insides of a human body. Immediately, a number of scientists launched themselves into this new field of research. Only a few months later, by pure chance, the French physicist Henri Becquerel observed that natural uranium salts emit very similar rays without any need for outside intervention, but his observation went almost unnoticed at first.

Going back to the Curies, while Pierre had already obtained his doctorate in March 1895 for his work on magnetism, Marie Curie's scientific activity had come to a standstill; her daughter Irène was born in September 1897. Shortly after the birth, the urge to go back to research won out and by December she was continuing her investigations on "Becquerel radiation" as Becquerel's doctoral student. It seems unlikely that Becquerel had any idea of the significance of the radiation he had discovered, otherwise he would hardly have entrusted such an important subject to a woman.

Marie's workplace was now a shed where it was either unbearably hot or freezing cold, and water dripped onto the work tables through its leaking glass roof whenever it rained. This is where an important part of scientific history was written over the following years. Little Irène and the household were taken care of, quite contrary to the customs of the time, by her recently widowed father-in-law.

Many scientists were working with *artificially* produced radiation at that time. Marie Curie, however, was the only one to systematically examine various uranium-containing compounds as *natural* sources of radiation. Here, the piezoelectric electrometer developed by Pierre Curie and his brother provided her with valuable assistance, as it could very precisely measure the changes in electrical conductivity in the air caused by the radiation. Within a short time, Marie Curie made significant discoveries:

- The experiments confirmed that uranium-containing materials were not somehow "charged up" by the effects of sunlight, only to re-emit the radiation again later on; the cause of the radiation was the atoms themselves. But how could this be? Emitted radiation had to leave the atom changed in some way, but scientists who believed in the existence of atoms (there were other explanations at the time) considered them unchangeable. Marie Curie's hypothesis was the first decisive step towards refuting the assumption that atoms are unchangeable and indivisible.
- In her systematic search for other substances that emit radiation, she discovered the radioactivity of the element thorium. The fact that the

German chemist Gerhard Carl Schmidt had beaten her to this discovery by two months only became known later.

- When she examined individual fractions of a thorium-bearing mineral, she found that it had to contain another radiating element in addition to thorium.
- When she extracted the uranium content from pitchblende, another uranium-containing compound, she was surprised to find that the residue radiated more strongly than the pure uranium. It was clear that pitchblende, a slag-like waste product from mining that is composed of up to 30 different elements, had to conceal a previously unknown radiating element.

The assumption that atoms are not immutable spheres was sensational. In April 1898, Curie's first results were presented to the *Académie des Sciences* by her doctoral supervisor Gabriel Lippmann. She herself was not allowed to give the lecture, as women were not admitted as members of the Academy. In the script for this speech, Marie Curie introduced the term "radioactivity" for the first time.

A race began for the first description of the two predicted elements—whoever first succeeded in producing them in pure form and determining their physical properties would go down in history as their discoverer. Pierre Curie decided to abandon his own work on crystals and work with his wife on these naturally radiating materials. The couple worked hand in hand: Marie perfected the chemical separation of the material samples, while Pierre investigated the physical properties of the separation products.

* * *

In painstakingly detailed work, the Curies separated several tonnes of pitchblende into its components, always on the lookout for other components that might be responsible for the radioactivity of the raw material, in addition to the uranium. After many weeks, they had isolated tiny amounts of two elements. One of them resembled bismuth in its chemical properties. In July 1898, the couple called it polonium in honour of Marie Curie's homeland. In the same year, they also announced the existence of the second element they had discovered, which resembled barium. Although they did not succeed in obtaining it in chemically pure form and were therefore unable as yet to make any statements about its properties, they gave it the name "radium".

Their discovery that radium was actually created by radiating uranium was a sensation. The chemical elements had been considered as the immutable building blocks of the universe; that one element could become another had previously been unimaginable. The Curies continued to work hard to obtain radium in its pure form. Only then would its discovery be official. In 1902, they succeed in producing radium chloride, but it was not until 1910 that Mare Curie, together with André Louis Debierne, was able to isolate radium in its elemental form by electrolysis of a radium chloride solution.

Between 1898 and 1902, the Curies published a total of 32 scientific papers, some together and some apart. Pierre Curie's search for a better paid job nevertheless remained unsuccessful, because his curriculum vitae with the very late doctorate did not fit the expectations of the academic world. So their financial circumstances remained strained. Nevertheless, they publicised their process for the production of radium chloride in detail. It turned out that exposure to radium destroyed diseased, tumour-forming cells more quickly than healthy one, but they refrained from applying for patents. They shared their conviction that the fruits of scientific activity should benefit all humankind with Konrad Röntgen, who had also refrained from patenting his invention of X-rays.

* * *

The year 1903 would be a year of triumph for the Curies. In June, Marie Curie submitted her doctoral thesis on radioactive substances and was awarded her doctorate. In November, the couple was awarded the Davy Medal of London's Royal Society, England's highest honour in the field of chemistry. And to top it all, Marie and Pierre Curie, together with Henri Becquerel, received the Nobel Prize in Physics in December.

Originally, the French Academy of Sciences had only proposed Henri Becquerel and Pierre Curie to the Nobel Prize Committee, but not Marie Curie. However, some scientists found this unacceptable, including the Swedish mathematician Gösta Mittag-Leffler, who had also supported Sofia Kovalevskaya. As a member of the Nobel Prize Committee, he and Pierre Curie managed to get Marie Curie included on the list after all. The chemists on the Nobel Committee even pursued the strategy of not explicitly mentioning the discovery of the elements polonium and radium in the justification for awarding the prize. So the Curies accepted the Nobel Prize in Physics for their joint work on Becquerel radiation. The distribution of the prize, however, brought the proportions out of balance again: Becquerel received half of the prize money, the Curies the other half.

For Pierre Curie, 1903 was the high point of his scientific work. By June 1904, he had published 25 papers on radiation. This productivity now came to an abrupt end; in the following two years he published no further work. There were several reasons why he was unable to repeat his earlier successes:

- He was finally appointed professor at the Sorbonne. Because the well-paid position created especially for him was linked to a teaching position, he now had less time for work in the laboratory.
- The Nobel Prize brought a lot of disruption into the Curies' lives; they were constantly disturbed by journalists and photographers, at work and even at home. The interruptions were particularly upsetting for Pierre Curie.
- The unprotected handling of radioactive materials began to have consequences. The couple had not travelled to Stockholm in December 1903 for the Nobel award ceremony because Pierre had felt too ill and too busy to travel.

Marie Curie, on the other hand, intensified her work. When Pierre became a professor, a position had also been found for her: she was now responsible for the experiments as the official laboratory director. In August 1903, she suffered a miscarriage in the fifth month of her pregnancy, and in December 1904, Ève, the Curies' youngest daughter, was born. After a short break, Marie Curie resumed her research and also her teaching at the École Normale Supérieure. She had been lecturing there since 1897 and had been appointed as the first female faculty member in 1900.

* * *

Pierre Curie did not die from the effects of intense radiation, but rather from an accident at the age of only 46. On 19 April 1906, he went out to visit his publisher, but due to a strike by the employees, he had to return. On the way back, it was raining. His open umbrella may have obstructed his view, but somehow he failed to notice a heavily loaded horse-drawn cart. Pierre Curie was run over and died instantly. Marie Curie was devastated and did what she always did in times of crisis: she threw herself into her work. Her father-in-law continued to look after Irène, who was eight years old at the time, and Ève, who was just one and a half years old.

Four weeks after Pierre Curie's death, on 13 May 1906, the Sorbonne offered his orphaned chair to his widow. Marie Curie thus became the first woman to receive a teaching position at a French university. However, she was only hired as an associate professor; it was not until 1908 that she was appointed a full professor.

In February 1910, another death shook the family. Marie's father-in-law died. Once again, Marie Curie disappeared into her laboratory. Irène and Ève, who had been raised by their grandfather, later reported that their mother had been more interested in science than in her children.

It was in 1910 that Marie Curie made further ground-breaking discoveries:

- On the front page of the *London Times* of 9 August 1906, there was a letter to the editor by the famous physicist Lord Kelvin in which he advanced the theory that radium was not an element but a compound of lead and five helium atoms. Marie Curie once again intensified her efforts and, with the help of her colleague André Debierne, succeeded in isolating the element radium in pure form and determining its atomic weight. This proved beyond doubt that radium was indeed an element.
- She published her comprehensive textbook "A Treatment of Radioactivity."[3]
- She defined the amount of radiation of 1 g of radium^{226} as the international standard for radium emissions. It was now possible, for example, to standardise medical applications. Its unit of measure was accepted by the international scientific community and called the curie (physical units named after scientists always begin with a lower-case letter). In 1985, this unit was replaced by the SI unit, called the becquerel, which indicates the average number of atoms that decay per second.

* * *

Despite Marie Curie's undeniable successes, few people would accept that a woman could be a professor and make outstanding scientific achievements. When a seat became vacant in the *Académie des Sciences* in 1910 due to the death of a member, a public dispute erupted over whether Marie Curie should be admitted to France's scientific elite. False claims such as that she was Jewish and a foreigner resulted in another scientist being elected by a narrow majority as a new member of the prestigious society in 1911. It was not until over half a century later, in 1962, that Marguerite Perey, a doctoral student of Curie's, became the first woman to be elected as a corresponding member of the *Académie*. And it took another 17 years before a woman was accepted as a full member in 1979.

Marie Curie had barely recovered from the controversy when the now 44-year-old widow had to endure another mudslinging campaign in 1911. A few years after her husband's death, she had entered into a relationship with the physicist Paul Langevin, who was five years her junior. The latter was a student of Pierre Curie, was married, and had four children; the Langevin and Curie families were friends and also went on holiday together. For their meetings, Curie and Langevin rented a two-room flat. In the spring of 1911,

---

[3] Marie Curie, *Traité de Radioactivité*. 2 volumes. Paris: Gauthier-Villars (1910); German edition: *Die Radioaktivität*. Leipzig: Akademische Verlagsgesellschaft (1911–1912), translated by B. Finkelstein.

the lovers knew that their relationship had been discovered, because incriminating letters had disappeared from this flat. From then on, they had to fear that their relationship would be made public. In mid-1911, Langevin's wife filed for divorce, and from November of that year, rather hostile newspaper articles appeared almost daily, followed soon afterwards by excerpts from the letters. In the eyes of the public, Curie was the guilty party, having led an honourable family man astray. Again, it was rumoured that Curie was Jewish, which for many of her contemporaries would have constituted an explanation for what had happened. The hunt culminated in the speculation that Curie's affair with Langevin had begun while Pierre Curie was still alive and had thus driven him to suicide.

Only a few people dared to take her side. One of them was Albert Einstein, who wrote to her at the end of November:

> (...) I am so incensed by the scurrilous way in which the public currently dares to engage with you (...) I feel compelled to tell you how much I admire your intellect, your drive and your honesty, and that I consider myself lucky to have met you in person in Brussels. If the rabble continues to engage with you, just don't read this nonsense, but leave it to the reptile for which it was fabricated (...).[4]

Einstein and Curie became lifelong friends.

At the very time that Marie Curie was being socially annihilated in France, a telegram arrived from the Nobel Prize Committee announcing that she would be awarded the Nobel Prize in Chemistry in recognition of her discovery of the new elements polonium and radium. Although she was advised not to attend the award ceremony in person, she travelled to Stockholm accompanied by her sister Bronya and her daughter Irène.

The award ceremony on 10 December 1911 was the crowning glory of a disastrous year for Marie Curie. But the stress of the past months had taken its toll. The depression from which she had suffered to varying degrees since childhood intensified again, in addition to acute kidney problems. In January 1912, she spent several weeks incognito in a private clinic. In March, she underwent a kidney operation and recuperated for months in a house near Paris that she had rented under her birth name Sklodowska.

She stayed away from her Paris laboratory for a whole year and only returned in December 1912. In the meantime, the scandal had fizzled out, the Langevin couple had separated, and Marie Curie was not mentioned by

---

[4] Albert Einstein to Marie Curie on 23 November 1911, from: Walter Isaacson, *Einstein: His Life and Universe*. Simon & Schuster (2008).

name in the divorce settlement. Marie Curie and Paul Langevin's affection for one another did not survive the scandal and there were no more love affairs in Curie's life. It is an irony of fate that, two generations later, Curie's granddaughter Hélène and Langevin's grandson Michel married.

* * *

Since 1909, the construction of an *Institut du radium* had been planned for research on radioactivity, and in 1912, construction work finally began in Paris in the rue Pierre-Curie. The department for fundamental research was to be run by Marie Curie, while a second department was planned for work on medical applications of radium. Shortly afterwards, Marie Curie took charge of the completed building, just as the First World War broke out. Marie Curie acted immediately and, as a Nobel laureate, managed to convince wealthy people to donate money for small vehicles equipped with X-ray machines, which could be used to examine and treat wounded soldiers directly at the front. As early as November 1914, the first two vehicles were in the combat zones. The crew of one of these mobile X-ray trucks was none other than Marie Curie herself, her scientifically accomplished 17-year-old daughter Irène, and a military doctor. A total of twenty *Petite Curies* were built, and in parallel hundreds of X-ray stations were established, in which a total of 1.2 million people were examined during the war years—this was the breakthrough of X-ray technology. Marie Curie's initiative saved the lives of many thousands of soldiers in the country that had ostracised her as a foreigner only a few years previously. She summarised her experiences in her book "Radiology in War."[5]

In 1919, Marie Curie returned to the *Institut du radium* and made it a world class centre for radiological research. Recognising the trend for science to become increasingly specialised, she organised small teams of researchers to pursue their tasks independently. Between 1919 and 1934, the scientists at the Radium Institute published 483 papers, while 31 papers and books were by Curie herself. Marie was the very heart and soul of the institute. She took part in the work of each individual and was constantly surrounded by collaborators. Her work shifted ever more from research to organisational tasks. But until the end of her life, Marie Curie continued to work on the isolation, concentration, and purification of polonium and actinium, another radioactive element.

Thanks to its connections all over the world and cooperation with industry, the *Institut du radium* was comparatively well equipped and well endowed

---

[5] Marie Curie, *La Radiologie et la Guerre*. Paris: Félix Alcan (1921).

with the precious radium, its main research material. Only a few other institutions could compete at this level: the Cavendish Laboratory in Cambridge, England, the Kaiser Wilhelm Institute for Chemistry in Berlin, and the Radium Institute in Vienna.

In addition to her many tasks, Marie Curie found the time to work on the Commission for Intellectual Cooperation of the League of Nations, the forerunner of UNESCO. In this capacity, she championed scientific standards that are taken for granted today:

- the compilation of an international bibliography of scientific papers,
- the development of standards for international scientific scholarships,
- the protection of intellectual property rights of researchers regarding their discoveries.

Her daughter Irène and her son-in-law Frédéric Joliot became stars of the Radium Institute over the years. Curie witnessed their triumphant discovery of artificial radioactivity in early 1934. But by the time the Joliot–Curie couple were jointly awarded the Nobel Prize in Chemistry in December 1935 for proving that radioactivity could also be produced artificially, Marie Curie had already lost the battle against radiation sickness.

* * *

From the very beginning, Marie and Pierre Curie, like their colleagues, handled radioactive substances with no attempt to protect themselves from the radiation. In 1901, Pierre Curie and Henri Becquerel had even listed the radiation damage they had observed.[6] They reported with scientific meticulousness on the painful skin burns, inflammations, and detachments, but they clearly believed that their research was worth these sacrifices. Today, it is difficult to understand why it was overlooked for so long that radioactivity also causes long-term and irreversible damage to health. Until the late 1920s, radium compounds were even considered beneficial to health, and by the mid-1930s they were still popular as additives to cosmetics and luxury foods. From toothpaste to the glowing eyes of stuffed animals at night, the possible uses of radium seemed unlimited.

Marie Curie's fingertips were already badly damaged by 1898. When Pierre Curie gave a lecture to the prestigious *Royal Institution* in London in June 1903, his hands were already so sore that he could barely change his

---

[6] Pierre Curie and Henri Becquerel, Action physiologique des rayons du radium, *C.R.T (Cathode Ray Tube)*, 132 (1901) pp. 1289–1291.

clothes and spilled some of the material while demonstrating the properties of radium. From the early 1920s, a cataract caused by exposure to radiation affected Marie Curie's vision in both eyes, so that her writing became increasingly large and she had to be led around by her daughters. Her health continued to decline and more and more often she was not strong enough to go to the laboratory. In the Easter holidays of 1934, she made one last trip with her sister Bronya.

Marie Curie died of leukaemia on 4 July 1934. Her personal notebooks from the 1890s are now considered too dangerous to handle because of their radioactive contamination and are kept in lead-lined boxes. Even her private cookbooks have to be kept in a radiation-proof container.

Probably the most famous female scientist in history, Marie Curie became an icon throughout the scientific world. In 1995, she became the first woman ever to be enshrined in the Panthéon in Paris, together with her husband. According to the 2009 *New Scientist* survey, she is the "most inspiring woman in science".

10

Lise Meitner (1878–1968): Discoverer of Nuclear Fission

Marie Curie is the most famous natural scientist in world history and her name is inextricably linked with the history of radioactivity. By contrast, the name of another woman who made an equally significant contribution to the understanding of this field of physics at almost the same time remains virtually unknown to the general public today. This was Lise Meitner.

In the 1920s and 1930s, she was widely recognised as an outstanding experimenter and theorist; Albert Einstein called her "our German Marie Curie". Her discovery of the radioactive isotope protactinium-231 in 1917 was already worthy of a Nobel Prize. She was nominated for this highest award a total of 48 times, but it always went to someone else. Even when she was the first to succeed in unifying the sometimes contradictory results of scientists from several different countries, realising that atomic nuclei can be split, releasing a small part of their mass in the form of energy, she was denied recognition for this achievement. Indeed, only a few months before she was able to put together the final pieces of the puzzle, resulting from many years of research, Lise Meitner had had to flee Germany overnight. From abroad, she remained in contact with her long-time colleague and friend Otto Hahn, who completed their joint life's work in Berlin—and the Nobel Prize went to him alone.

The fact that Lise Meitner opened the way to using nuclear energy and thus changed the world was never really appreciated in her own time. Today, many schools bear her name, but she is still largely overshadowed by her fellow physicist Marie Curie.

L. Jaeger, *Women of Genius in Science*, https://doi.org/10.1007/978-3-031-23926-7_10

* * *

Elise Meitner, known as Lise, was born in Vienna on 7 November 1878. Her mother Hedwig and her father, the lawyer and unflinching freethinker Philipp Meitner, had moved far away from their Jewish roots. Religion hardly played any role in the Meitner household. Some biographies report that Lise and her seven siblings were baptised and brought up as Protestants, but they all converted to Christianity only as adults. Their parents' tolerant attitude and commitment to modernity can also be seen in the fact that Lise and her four sisters were among the first young girls in Vienna who did not have to wear a laced corset.

Even as a child, Lise Meitner was interested in natural phenomena. For example, she wanted to find out why a film of oil on a puddle of water would shimmer colourfully. And at the age of eight, she slept with her mathematics book under her pillow. But there was no room in the school system for inquisitive girls: in 1892, Lise was 14 years old and she found she had reached the end of the school career for young ladies. Taking the Abitur was possible in principle with the help of private tutors, but it was anything but common. Attending university was still impossible at this point. But that was nevertheless Lise Meitner's goal:

> I was obsessed with the desire to prepare for the Gymnasium Matura since my thirteenth year, in order to study mathematics and physics (...)[1]

In 1897, the rules for women changed; now they could also study in Vienna. Lise's eldest sister Gisela led the way. She convinced her father to pay for private lessons to prepare her for the Matura, and in 1900 she began studying medicine at the age of 24. Lise was two years younger than Gisela and wanted to follow in her sister's footsteps. But her parents insisted that she take the French teacher's exam. Only later, when Lise was already 20 years old, did her father also allow her private lessons. In just two years, she had caught up with four years of grammar school education and passed the Abitur examination at a Viennese grammar school as an external student in 1901. Only four out of fourteen girls passed the exam that year, including Henriette Boltzmann, the daughter of the physicist Ludwig Boltzmann. In the same year, Lise Meitner began to study physics and mathematics, and she also took philosophy classes, as these were still part of the compulsory programme at that time. She had wasted valuable years, but now the 22-year-old's lifelong dream finally came true.

---

[1] Anne Hardy and Lore Sexl, *Lise Meitner* (German edition). Rowohlt (2002), p. 18.

* * *

Lise Meitner's most important teacher at Vienna University was the great Ludwig Boltzmann, who had revolutionised thermodynamics. In his lectures, he also dealt with ethical questions, but he did not talk about the brand new and much discussed quantum theory, nor about the latest experiments on radioactivity. In February 1906, Lise Meitner completed her doctorate on a rather classical topic: "Heat conduction in inhomogeneous bodies". It was her doctoral supervisor Franz Exner, a friend of Konrad Röntgen, who familiarised her with the state of knowledge in radiation physics at that time. One of the most important advances had been made at the turn of the century by Ernest Rutherford when he distinguished three types of radiation according to their physical properties:

- Alpha radiation (or "Becquerel radiation") has a range of about 10 cm in air, and even a sheet of paper can shield from it. A few years later, it was discovered that it consisted of packets of two protons and two neutrons—alpha rays were therefore nothing more than a stream of helium nuclei flying through space. Unstable isotopes were identified as the source, emitting these particles during spontaneous transitions to a stable state. In addition to uranium, the elements polonium and radium discovered by the Curie couple were also natural alpha emitters.
- Beta radiation has a range of about eight metres in air and a sheet of aluminium or iron several millimetres thick is needed to protect oneself from this radiation. At the time of Meitner's work in Vienna, it was not yet known that these were electrons, which are released when a neutron is converted into a proton, or positrons, which are produced during the reverse conversion.
- Gamma radiation can still be measured several hundred metres away from the source and, unlike alpha and beta radiation, cannot be deflected by magnetic fields. It takes lead plates, a thick layer of reinforced concrete, or even water to stop them. They are particularly energetic and, like light and X-rays, electromagnetic in nature.

Lise Meitner was fascinated by this subject and, immediately after her doctorate, the now 28-year-old applied to work with Marie Curie in Paris. She was turned down and initially stayed in Vienna, where she did experimental work at the Institute for Theoretical Physics. She let parallel alpha rays collide with foils of different metals and found that the higher the atomic mass of the foil metal, the greater the observed scattering. This finding was so significant that it was published in 1906 and 1907 in a leading physics journal.[2] Indeed, it inspired Ernest Rutherford to conduct his own experiments, which led to his atomic model with an atomic nucleus and orbiting electrons.

These initial successes encouraged Meitner to move to the Friedrich Wilhelm University in Berlin in 1907. Women were not yet allowed to enter this world class centre for theoretical physics, yet she quickly found supporters. Even Max Planck, the founding father of quantum physics, who was in fact opposed to admitting women to universities, allowed her to attend his lectures and invited her to his home. However, it was another encounter in Berlin that was to give Lise Meitner's life its decisive direction and transform the originally intended one or two years in Berlin into a stay of over three decades.

* * *

In her search for a laboratory in which she could experiment, the physicist Meitner met the chemist Otto Hahn, who was almost her age and had come to Berlin a year earlier looking for an assistant. Working on the topic of radioactivity, he would have been something of an outsider in the university chemistry department, so he had set up his laboratory in the physics department. Meitner's and Hahn's decision to work together was a stroke of luck in the history of science, because the two complemented each other perfectly: Hahn dealt systematically and methodically with the detection of new elements or isotopes, but when it came to bold and creative explanations for the observed radiation phenomena, it was Meitner who came into her own. When Otto Hahn once tried to come up with a theory, she simply said:

Hähnchen, let me do it, you don't know anything about physics.

In her first year, cooperation was still quite awkward, because many members of the institute would have considered the presence of a woman as a disturbance. For this reason, Lise Meitner was only allowed to enter and leave the laboratory in the basement via a side entrance leading directly to the street. All other rooms and corridors of the building were off-limits to her. Even a visit to the toilet was out of the question, because there was no ladies' room. But this situation did not last long; by 1908, Lise Meitner was officially allowed to move around the institute, as women were now allowed

---

[2] Lise Meitner, Über die Absorption der Alpha- und Beta-Strahlen, *Physikalische Zeitschrift*, 7 (1906), pp. 588–590; Lise Meitner, Über die Zerstreuung der α-Strahlen, *Physikalische Zeitschrift*, 8 (1907), pp. 489–496.

to study in Prussia too. But the discrimination continued. It is reported that some of her colleagues ignored Lise Meitner and only greeted Otto Hahn when they met the two of them together. Despite all the adversities, Meitner later described her years in the physics department at the University of Berlin as her happiest:

We were young, merry and carefree (...)[3]

One of the most important discoveries from these first years of collaboration with Hahn was the possibility of determining the radioactive source material from the energy content of the relatively massive alpha particles and even discovering completely new radioactive substances. They mistakenly called the phenomenon "radioactive recoil"; only much later did they realise that radioactive atoms can decay and their fragments absorb certain amounts of kinetic energy in the process. When Meitner and Hahn carried out the first tests with this method, they quickly discovered two new radioactive isotopes.

In 1910, Hahn became a full professor, while Meitner remained without a salary and without an official function. In 1912, the Meitner–Hahn team moved from Berlin University to the newly founded Kaiser Wilhelm Society, now known as the Max Planck Society. Hahn would contribute to the establishment of an institute of chemistry, in which he set up and headed a department of radiochemistry as a counterpart to Marie Curie's laboratory in Paris. Meitner and Hahn had previously conducted their research with equal responsibilities, but in keeping with the spirit of the times, only Otto Hahn was considered for the position, which carried an annual salary of 5000 marks; Lise Meitner continued to work in his new department with guest status and without a salary. It was not until a few months later, in the winter of 1913/14, that Max Planck found her a position as an assistant at his Institute for Theoretical Physics at the Friedrich Wilhelm University. This was because, after the death of her father in 1910, Meitner's financial situation would have been even more strained than ever and he feared she would have to return to her family in Vienna. It was not only the first paid position for the 35-year-old Lise Meitner personally, but the first ever endowed position for a female scientific assistant in Prussia. Meitner celebrated this unique success with a dinner party at the Hotel Adlon.

Shortly afterwards, Planck had Lise Meitner's annual salary doubled to 3000 marks, so that he could dissuade her from accepting a lectureship offered to her in Prague. Otto Hahn's salary was still considerably higher than Meitner's, but the time of financial worries was over. In 1914 alone, Hahn

---

[3] Charlotte Kerner, *Lise, Atomic Physicist—The Life Story of Lise Meitner*. Beltz und Gelberg (1996).

received 66,000 marks in royalties from the sale of the isotope radium-228, also known as "mesothorium" or "German radium," which Hahn had discovered before working with Meitner and which was used in the fight against cancer. The ten percent of this sum that he gave to Meitner corresponded to twice the annual salary of his "assistant."

* * *

Like Marie Curie in Paris, Lise Meitner was also torn from her work by the First World War. Initially, she had been able to continue working in Berlin, but in July 1915 she was called up for active service by her home country, Austria-Hungary. Like Marie Curie on the opposite side, she worked as an X-ray technician, first deployed to the Polish front, and later to the Italian front. Soon there was not much left of her initial enthusiasm for the war. After her discharge in September 1916, she returned to Berlin.

Otto Hahn also arrived back in Berlin at almost the same time. He had taken part in the World War in a completely different capacity: as a member of the group run by Fritz Haber, co-inventor of artificial fertiliser and founding director of the Kaiser Wilhelm Institute for Physical Chemistry, he was involved in the development of chemical warfare. On 22 April 1915, Hahn personally supervised the first use of chlorine gas as a weapon of war in Flanders. The cloud of poisonous gas surprised the soldiers in the trenches and killed about 5000 people, while another 10,000 were so badly burned that they were permanently incapacitated. The impressions Hahn gained at the front made him doubt his actions. Meitner had also imagined the war to be less brutal, but she had nevertheless written to Hahn: "I congratulate you on the fine success at Ypres."

From 1917, Meitner and Hahn worked together again in Berlin and immediately discovered the new element protactinium-231. In the period that followed, Lise Meitner made several major steps forward in her career:

- In 1917, she set up a physical radioactivity division at the Kaiser Wilhelm Institute for Chemistry, tailored to her needs, five years after Otto Hahn had designed his chemical radioactivity division.
- In 1918, she became head of this department and had her own research funds. Meitner and Hahn continued their fruitful exchange, but after 1919 there were no further joint publications for a long time.
- In 1922, Lise Meitner became the first female physicist to habilitate at a university in Germany. Because this title was not associated with an official teaching position, she taught as a private lecturer.

- In 1926, she was the first woman in Germany to be appointed associate professor for experimental nuclear physics. In this position she still did not have the status of a civil servant, but was allowed to supervise doctoral students. By 1933, three students had received their doctorates from her.

It was still most unusual for a woman to be a professor. The following anecdote shows how entrenched people's prejudices were. The title of her inaugural lecture in 1922 was: "The significance of radioactivity for cosmic processes." Unthinkingly, one newspaper described how Lise Meitner had lectured on "cosmetic processes."

* * *

By this time, Lise Meitner had probably come further than her wildest dreams. But then her scientific career was suddenly interrupted. In March 1933, the National Socialists came to power in Germany and within a very short time passed a law forcing Jews out of the civil service. However, this "Law for the Restoration of the Professional Civil Service" could not be applied to Meitner for two reasons: firstly, an additional paragraph exempted anyone who had served in the World War from this regulation, and secondly, she was Austrian. Nevertheless, in April 1933, her authorisation to teach at the university was withdrawn; she was no longer an associate professor. In order to make this process compatible with the law, she was accused of not having served at the front during the war. Otto Hahn declared his solidarity with her and all other affected colleagues at the university and also left the University of Berlin in January 1934. Not much changed for Meitner and Hahn; their scientific centre was and remained the Kaiser Wilhelm Institute. This was not a state-funded institution, but one that formed a bridge between politics and industry. Because the focus there was less on ideology than on profitability, Lise Meitner was able to continue working in her department for several more years. Carl Bosch, director of IG Farben and thus the main sponsor of the Kaiser Wilhelm Institute for Chemistry, promised Meitner that she could keep her position at the institute permanently. But Meitner enjoyed little protection outside the institute.

During this difficult time, Lise Meitner convinced her closest colleague and most valued friend Otto Hahn to work as a team again after the fourteen-year break. Together with the younger Fritz Straßmann (1902–1980), they set about further researching the atomic nucleus. For in the world of radiation physics, the 1930s were of decisive importance.

- In 1932, the neutron was discovered.

- In 1934, the Italian Enrico Fermi bombarded uranium with neutrons for the first time. Meitner, Hahn, and Straßmann also tried this experiment. For four years, Hahn and Straßmann tried to isolate and determine the resulting radioactive substances, while Meitner searched for theoretical explanations.
- In 1937, Marie Curie's daughter Irène Joliot-Curie and the Serbian physicist Paul Savitch also tried to work out what was happening when uranium was bombarded with neutrons.

The best experimentalists and chemists in the world worked flat out to gain a better understanding of nuclear physics with the help of the uranium–neutron experiment. Fermi had assumed that bombardment with neutrons would yield so-called transuranium elements, i.e., ones with higher atomic weight than uranium (92 protons). Others thought they had found an isotope of radium (88 protons). For years, no one succeeded in separating the mixture of radioactive products and identifying the individual components.

Meitner, Hahn, and Straßmann came to the conclusion that the prevailing interpretation of the experiments, especially the alleged discovery of radium isotopes, could not be correct. Hahn refined his chemical methods with Straßmann, while Meitner devised new experiments to shed more light on individual reaction processes. With better analytical possibilities than ever before, they decided to repeat the uranium–neutron experiment. It later turned out that it was precisely this experiment that brought the breakthrough. But Lise Meitner was not on site for this decisive experiment.

* * *

For five years, Meitner had managed to make herself virtually invisible in a society that was increasingly hostile to Jews. This changed abruptly with the Anschluss of Austria in March 1938. She was now German and lost the privileges that applied to foreigners. Meitner hesitated for valuable months before going into exile. When she finally decided to do so, the possibilities for leaving the country had been further restricted and her application to leave was rejected. Lise Meitner was now in acute danger. When she entered the institute on the morning of 12 July 1938, Otto Hahn informed her of the escape plan he and some foreign friends had forged: the very next morning she was to take the train to Holland. In order not to arouse suspicion, she stayed at the institute until 8 p.m. as usual, correcting a paper by a staff member that was intended for publication. Hahn helped her pack two small suitcases and gave her a diamond ring he had inherited from his mother in case of emergency. Meitner spent the night at Hahn's house. The next morning, she met the Dutch chemist Dirk Coster at the railway station as if by chance, and they pretended they were both on their way to a conference. Together they reached Holland without incident. A few weeks later, Meitner travelled on to Stockholm, where friends had organised a job for her. It was a major blow for the 59-year-old physicist, who had by then become accustomed to perfect research conditions: it was not possible to conduct world-class experiments in her new laboratory, which was only very poorly equipped. Indeed, the simplest equipment was lacking.

It was all the more important for her that Otto Hahn and Fritz Straßmann continued to keep in touch at great risk. Secretly, they sent her their latest research results by letter, hoping that she could provide the corresponding explanations.

In November 1938, Lise Meitner and her nephew Otto Robert Frisch, who was also a well-known physicist, met in Copenhagen with Otto Hahn and Niels Bohr to discuss the experimental results obtained when uranium was bombarded with neutrons. Meitner reiterated her view that the reaction product could not be radium.

Back in Berlin, Otto Hahn and Fritz Straßmann intensified their experiments and found that some of the alleged radium isotopes behaved chemically like barium, an element with only slightly more than half the atomic mass of uranium. At the end of December 1938, Hahn wrote to Meitner:

> Maybe you can suggest some fantastic explanation. We know ourselves that it [uranium] cannot actually decay into Ba (...) Would it be possible for uranium 239 to break into a Ba and an Ma? Of course, I would be very interested to hear your opinion. Perhaps you could calculate something and publish it.

By "Ba" he meant barium, by "Ma" the element technetium, which was still called masurium at the time. These few words reflect the great uncertainty about what was physically possible and what was not. Meitner was in charge of this topic. She wrote back immediately:

> For the time being, it seems to me that the assumption of such a far-reaching decay would be very difficult, but we have had so many surprises in nuclear physics that it can never be said without further ado: this is impossible.

Within a few days, she and her nephew Otto Frisch worked out a theoretical model of nuclear fission. In accordance with the droplet model of the atomic nucleus that prevailed at the time, they imagined the uranium nucleus to be an unstable oscillating droplet that tended to split in two when disturbed by a neutron. According to the number of protons present in the uranium atom, the other element had to be, not technetium (43 protons), but krypton (36 protons). But how to explain the very high speed with which the two fission droplets were driven apart? They had measured energies totalling about 200 meV per uranium nucleus. That was about 400 times the rest energy of an electron. Where could this energy come from? And why were the products of nuclear fission lighter than the original uranium nucleus by about one fifth of the mass of a proton (200 meV)?

Lise Meitner was the first person to put the pieces of the puzzle together. Hahn had told her how mass was missing in the uranium–neutron experiment after the bombardment. She now calculated the energy equivalent of the mass defect according to Albert Einstein's famous equation $E = mc^2$, and came up with exactly 200 meV. Meitner and Frisch substantiated this assumption in a simple experiment: they measured the "recoil" of the fission fragments with a Geiger counter and found that the measured energies corresponded exactly to the predictions.

With this, Meitner had not only explained the high energies of the fragments produced in the experiment, but also confirmed Einstein's famous formula experimentally for the first time. It was also immediately clear to her that the immense nuclear energy available offered sensational technological possibilities. For many years, scientists all over the world had been freely discussing the subject of radioactivity in a completely open manner. With the discovery of nuclear fission and the possibility of harnessing the immense energies lying hidden in atoms, international cooperation suddenly came to an end. As early as spring 1939, the German military showed interest in possible applications of nuclear fission. Just over six years later, the first atomic bombs were dropped.

* * *

Because it was impossible for Hahn and Straßmann in Germany to name the non-Aryan Lise Meitner as a co-author, the two published the proofs of nuclear fission without naming her as a co-author.[4] Only a few weeks later, Meitner and Frisch published their theory of nuclear fission in an English-language scientific journal.[5] This created the impression that the chemists in Germany had discovered fission, but the scientific community knew very well that Meitner and Hahn were equally involved in this success. Soon the term

---

[4] Otto Hahn and Fritz Straßmann, Über den Nachweis und das Verhalten der bei der Bestrahlung des Urans mittels Neutronen entstehenden Erdalkalimetalle, *Die Naturwissenschaften*, 27, 1 (1939), pp. 11–15.
[5] Lise Meitner and Otto Frisch, Products of the Fission of the Uranium Nucleus, *Nature*, 143 (1939), pp. 471–472.

"atomic fission" proposed by Lise Meitner and Otto Frisch was on everyone's lips.

Historians of science disagree over whether Hahn was bothered in the end about not sharing the fame. In 1938, he had still demonstrably tried to include Lise Meitner as an author on the publication. But after the war he claimed that he had not needed Meitner's inspiration to demonstrate nuclear fission in his chemical experiments. The fact is that Hahn alone was awarded the Nobel Prize for nuclear fission in 1944: this decision was not announced until 1945, and the prize was awarded in 1946. Lise Meitner, Otto Straßmann, and Otto Frisch were left empty-handed. Disappointed, Meitner wrote to her former colleague James Franck in December 1946, when Otto Hahn accepted the Nobel Prize:

> Since I am, after all, part of the past that is to be repressed, Hahn never mentioned our long-standing collaboration or even my name in any of the interviews where he spoke about his life's work.[6]

Nevertheless, Meitner remained close to the man with whom she had conducted research for three decades.

* * *

Despite much opposition, personal disappointments, and poor working conditions in her new home country, Lise Meitner enjoyed international recognition. In 1946, she travelled to the USA, where the public showed great interest in the "Woman of the Year 1946." A New York radio station broadcast an interview with Meitner and the First Lady of the USA, Eleanor Roosevelt. At Princeton, Harvard, and Columbia University, she lectured and discussed with Albert Einstein, Enrico Fermi, Edward Teller, Victor Weisskopf, and Leo Szilard. In Washington, DC, she spent an evening with the Nobel laureate James Chadwick, who, as the highest-ranking military officer, had led the British mission on the Manhattan Project. All these interlocutors had contributed to the creation of the atomic bomb, but Meitner had refused to participate in the construction of this weapon, despite repeated requests. By now she was a convinced pacifist and it would not have pleased her that she was now being called the "mother of the atomic bomb" by the Americans. Back in England, she met other greats of physics: Erwin Schrödinger, Wolfgang Pauli, and Max Born. Indeed, she maintained a lifelong friendship with Schrödinger and Pauli.

---

[6] Jost Lemmerich, *Lise Meitner on the Occasion of Her 125th Birthday*, exhibition catalogue of the Staatsbibliothek zu Berlin Preußischer Kulturbesitz (2003).

While strongly condemning the use of nuclear power, she supported its peaceful use. She was involved in the planning and construction of Sweden's first nuclear reactor R1 in Ågesta, which was approved in 1957 and commissioned in 1964, producing 10 megawatts of power. In the 1950s and 1960s, Meitner visited West Germany and spent several days with her long-time companion Otto Hahn and his family. Hahn wrote in his memoirs that he and Meitner remained close friends for life.

At the age of over 80, Lise Meitner went into partial retirement and moved to Great Britain, where most of her relatives had emigrated. She still gave lectures and worked part-time in the laboratory.

Lise Meitner died on 27 October 1968 at the age of 89, three months after Otto Hahn. One last great honour befell her 25 years later: in 1994, the competent international commission agreed that element 109, which had been artificially produced in Germany by fusing bismuth with iron ions, should be named Meitnerium.

11

Emmy Noether (1882–1935): The Most Important Mathematician of All Time

In the fifty years between 1900 and 1950, the laws of physics were rewritten. The further scientists advanced into the world of quantum physics, the more obvious it became that Newton's laws of classical physics only lead to useful results in the world of our daily experience, but not on microscopic scales. In order to be able to explain processes at the atomic level as well, the entire edifice of physics had to be rebuilt from scratch. Something similar happened almost in parallel and to a no less dramatic an extent in mathematics. Here, it was through the confrontation with infinities that many thousand-year-old certainties had to be thrown overboard and replaced by new mathematical relations before this subject could once again become consistent.

It was a woman who played a central role in this period of change in both sciences: Emmy Noether. Her ingenious and profound thoughts led to decisive innovations in both mathematics and theoretical physics:

- She devised new concepts that gave consistency to the theory of relativity and later provided particle physics with a tool for classifying the bewildering array of quantum particles.
- It gave physicists a scientific basis for their widespread belief in the unity of nature.

For further details about Emmy Noether, see also: Lars Jaeger, Emmy Noether—Ihr steiniger Weg an die Weltspitze der Mathematik, Südverlag, Konstanz (2022).

© The Author(s), under exclusive license to Springer Nature Switzerland AG 2023
L. Jaeger, *Women of Genius in Science*,
https://doi.org/10.1007/978-3-031-23926-7_11

- Her unique way of thinking took her into spheres of the highest level of mathematical abstraction, without which today's algebra and with it many other mathematical subfields would be inconceivable.

* * *

Emmy Noether was born on 23 March 1882 as the eldest daughter of a Jewish family in Erlangen, Franconia. Her father Max Noether was a well-known mathematics professor at Erlangen University. As a teenager, Emmy received the usual school education for a girl in a family of her social status, in which playing the piano was more important than learning arithmetic. In other ways, too, she was little different from other girls of her age; she was an enthusiastic dancer and loved attending balls and similar events. However, her mind was set, not on a quick marriage, but on further learning. The fact that she decided to train as a language teacher, just like Lise Meitner, was no coincidence. For a high school diploma was not intended for girls, and a high school diploma was not necessary to become a teacher. At the age of 18, Emmy Noether passed the Bavarian state examination and could thus have worked as a certified teacher of English and French at Bavarian girls' schools.

Here, there is another parallel with Lise Meitner: instead of working as a teacher, Noether turned to studying one of the natural sciences, despite the seemingly insurmountable obstacles. Officially, women were still forbidden to study in 1900, but at the University of Erlangen, where her father taught, she was allowed to attend lectures as a guest student. At the same time, she prepared for an external Abitur examination, which she passed in 1903 in Nuremberg at the Königliches Realgymnasium. It was precisely from this year that women were first granted access to universities in Bavaria.

Emmy Noether's decision to study mathematics may have had something to do with her brother Fritz, who was two years younger and mathematically gifted, and who later became a well-known exponent of the subject and a professor in Breslau. Perhaps Emmy wanted to show that she was at least her brother's equal. In 1907, she received her doctorate in Erlangen—two years before Fritz—with the highest grade.

Just like Lise Meitner a year earlier in Vienna, Emmy Noether had now reached the end of the line with her doctorate in Erlangen. Although it was possible for women—with considerable personal and financial commitment—to obtain a school-leaving certificate, a degree, and a doctorate, it was simply unthinkable that a woman could take up a paid position at an institute. Emmy Noether had no choice but to support her father Max Noether free of charge in his research at the Erlangen Mathematical Institute.

Even in her subordinate position as a merely tolerated assistant, her mathematical abilities attracted the attention of the professional world. Just one year after her doctorate, in 1908, she was elected to the Italian *Circolo Matematico di Palermo*. In 1909, the prestigious German Mathematical Association DMV accepted her into its ranks—even before her brother Fritz—and allowed her to be the first woman to give a lecture at its annual conference. Further honours and invitations to speak were to follow. It is a paradox typical of that time: Emmy Noether achieved significant success on the international stage as an outstanding, innovative mathematician, and at the same time it was generally accepted that she would lead a shadowy academic existence at Erlangen University, without position or salary.

* * *

In 1915, Emmy Noether attracted the attention of the two greatest mathematicians of her day, David Hilbert and Felix Klein. The two invited her to join them in Göttingen, which was a world class centre for mathematical research at the time. This was because Noether had become an expert in a field in which the two mathematical titans were unable to make any progress of their own: what was known as invariant theory.

The progressive David Hilbert even suggested that Noether should habilitate in this important and very topical field of mathematics. Although the Prussian state had forbidden universities to habilitate women, almost all the mathematicians in the relevant faculty at Göttingen signalled their approval, something hitherto unheard-of! Noether's habilitation attempt failed largely because of the categorical refusal of the philosophy professors in the same faculty. The faculty minutes from 1915 show that even the supporters of Emmy Noether's habilitation had difficulty with the idea of habilitating a woman. The eminent mathematician Edmund Landau wrote in his report:

> How easy it would be for us if it were a man with exactly the same work, lecturing skills, and serious aspirations. I would much prefer it if this expansion of our teaching programme could be made possible without the associated habilitation of a lady. (...) I have hitherto had the worst experiences with regard to female students as far as achievement is concerned, and consider the female brain unsuitable for mathematical production; Miss N, however, I consider to be one of the rare exceptions.[1]

---

[1] Landau's expert opinion of 1.8.1915 following his circular of 20.7.1915, Mathematical and Scientific Examinations Office, file "Prof. Noether"; available at https://www.cordula-tollmien.de/noethertollm ien1990Seite176.html.

Since Noether's first attempt at habilitation failed, she had to hold her mathematics lectures under Hilbert's name for four years. It was not until 1919, after the loss of the World War and the subsequent revaluation of the position of women in society, that she was habilitated and was able to teach as a private lecturer under her own name. However, there was still no talk of equality: Emmy Noether continued to work unpaid.

By this time, she had further developed her international reputation. Numerous papers in the most important German journal for mathematicians, the Mathematische Annalen, had raised her profile. Almost every year, she published her results in the DMV's annual report. And so, the "daughter of Max Noether" had left the shadow of her famous father. When the latter died in December 1921, he was described as the "father of Emmy Noether".

In 1918, even before her habilitation, Emmy Noether published her most important work for theoretical physics. The occasion was the problem that David Hilbert and Felix Klein had brought her to Göttingen to solve. Einstein had published his general theory of relativity (GR) in November 1915 with the collaboration of the Göttingen mathematicians, and ultimately also in competition with them. One open point in it, which was hotly debated by Hilbert and Klein, was that the principle of conservation of energy seemed to be violated in the equations of GR. It was like writing the trivial equation: "0 is equal to 0." Did this redundant statement mean that the law of conservation of energy had no meaning in GR, or that it was not even valid? Would Einstein's theory have to be discarded because of this? Or was there a specific reason for the strange behaviour of Einstein's equations?

Emmy Noether showed in quite a short time that the strange nature of the law of conservation of energy was peculiar to a whole class of theories which mathematicians and physicists refer to as *covariant*. She recognised that the conservation laws in all covariant theories, and GR was one of them, have this special structure. So energy was indeed conserved in GR.

This discovery is known today as Noether's second theorem. On the way to proving this theorem, she also proved a now much better known theorem. This established a general connection between invariance (symmetry) under a variable transformation in a physical equation and a correspondingly conserved variable, i.e., a physical variable that does not change in time. This first Noether theorem is much more central to today's physics than the second one about covariant systems, and the two are often lumped together and called *the* Noether theorem). Let us therefore look at it a little more closely.

- **Symmetry**: An object is said to be symmetrical if a certain manipulation leaves it unchanged. For example, if you place a cube on a table and rotate it through 90 degrees or a multiple of that, it looks the same as before. A sphere can even be rotated arbitrarily around any axis passing through its centre and it will always look the same as before. What is true for objects can also be true for equations. If the parameters in an equation are changed and the equation remains unaffected, mathematicians and physicists speak of symmetries or invariances. Newton's classical laws, for example, are invariant in position and time. This is because, when we calculate the trajectory of a ball falling from a tower, its position coordinates are irrelevant to the speed of the fall, the energy on impact, etc. Put another way, the ball falls in Berlin just as it does in Munich. Even a shift on the time axis leaves the laws of nature unchanged. It does not matter on which day the experiment is carried out. Invariances were Emmy Noether's speciality.
- **Conserved variables** are parameters in physics whose values always remain constant during physical processes. The law of conservation of energy, for example, asserts that, in a closed system, the total energy remains the same. In addition to energy, other important conserved variables are momentum, angular momentum, and electric charge.

Emmy Noether found an intriguing connection between symmetries and conserved quantities. Her first Noether theorem led to the following realisations:

- The fact that the equations of physics apply independently of *temporal* shifts is directly related to the law of conservation of energy.
- The fact that the equations of physics apply independently of *spatial* displacements is directly related to the law of conservation of momentum.
- The fact that the equations of physics apply independently of the *orientation of the system in space*, i.e., it makes no difference in which direction an arrow is shot, is directly related to the law of conservation of angular momentum.
- The fact that the equations of physics apply independently of the *phase* of an electrically charged particle is directly related to the charge conservation law.

This astonishing result was presented by Felix Klein to the Göttingen Mathematical Society in July 1918. Despite her obvious competence and reputation in the professional world, it was impossible for her, as a woman, to be a member of this society or even to give a guest lecture. But at least she was able to submit the publication in the Göttinger mathematische Nachrichten under her own name.[2]

Today it seems almost magical how quickly Noether found the solution and thus paved the way for the final triumph of GR. Hilbert and Klein were thrilled, and Albert Einstein from Berlin also expressed his admiration for Noether's work in a letter to David Hilbert in May 1918:

> Yesterday I received a very interesting paper from Miss Noether on invariants. It impresses me that such things can be understood from such a general point of view. It wouldn't have done the old guard at Göttingen any harm if they had taken some lessons from Miss Noether's school![3]

* * *

With the discovery of the special behaviour of covariant or symmetric equations in the known fundamental physical theories, Emmy Noether's had actually completed her task. But when she was working on the proof of the correlation between symmetry and the strange behaviour of the corresponding equations in certain cases, she came across an even more general relation. She discovered that for *every symmetry* found in a law of nature, there must be a specific conservation variable associated with it. The converse was also true: any physically conserved variable could be assigned an associated symmetry.

This general analogy between symmetries and conservation variables initially seemed to have no further significance for physics and mathematics. But a few decades after Emmy Noether's death, it became apparent that her theorem was one of the most important principles in all of physics. With it, Noether had given physicists a powerful tool to find their way in the world of quanta, which was proving to be increasingly complex. If they noticed an invariance in one of their equations, they could go in search of a corresponding conserved quantity. And vice versa, they could deduce symmetries from new conserved variables. In this way, new laws of nature were discovered in the 1960s and 1970s, along with decisive clues as to the fundamental structures of particle physics. Without Noether's theorem, it would hardly have been possible to develop today's standard model of elementary particles. Here are a few examples:

---

[2] Emmy Noether, *Invariant Variation Problems*, Nachrichten von der Gesellschaft der Wissenschaften zu Göttingen, Mathematisch-Physikalische Klasse (1918), pp. 235–257.

[3] Albert Einstein, *Collected Papers*, 9B, n. 548., Princeton University Press (1987), pp. 774–775.

- Thanks to Noether's theorem (applied to angular momentum in quantum mechanics), it was possible to explain the existence of two fundamentally different kinds of quantum particles called bosons and fermions.
- The fact that there are eight different gluons in the atomic nucleus, which give rise to the strong nuclear force, is a direct consequence of the fact that the underlying mathematical symmetry group is eight-dimensional.
- The three exchange particles of the weak nuclear force, the W + , W-, and Z particles, result from the three-dimensional structure of the associated mathematical symmetry group.
- In 1964, Noether's theorem explained why there must be a Higgs boson that gives all particles mass. This particle was discovered in 2012 as the last fundamental particle of our standard physical theory of the microcosm.

Noether's theorem also plays a central role in the development of new theories that are intended to go beyond the standard model. For example, it is an important criterion in the current development of a quantum theory of gravity, which aims to unite two previously incompatible theories: GR in the macrocosm and quantum theory in the microcosm.

Thus, from the second half of the twentieth century, Noether's theorem became one of the most important research tools in physics. The theoretical physicist and 2004 Nobel Prize winner Frank Wilczek from MIT formulated it like this:

This theorem has been a guiding star for 20th and 21st century physics.[4]

Even everyday physics can sometimes benefit from Noether's theorem. It is used in the simulation of waves on the surface of the sea or the flow of air over an aircraft wing, as well as to calculate the vibrations of bridges and the effects of atomic explosions.

Moreover, the connection found by Emmy Noether between symmetries in the basic mathematical equations of physics and the conservation variables in nature is probably the most beautiful and sublime principle of physics, because it gives an overall picture of the way the world looks.

* * *

The great significance of her theorem might have pleased Emmy Noether, but probably not interested her very much. For she was not interested in

---

[4] Emily Conover: *In Her Short Life, Mathematician Emmy Noether Changed the Face of Physics*, posted at www.sciencenews.org on 12 June 2018.

applications. She felt most comfortable in the deepest abstractions of mathematics. Her work between 1920 and 1935 was so far removed from any real world application that hardly anyone outside a very small circle understood what she was doing. And yet it is thanks to her work that the whole of mathematics could be placed on a new foundation during this period.

At the end of the nineteenth century, when mathematicians began to turn their attention to the concept of the infinity, they realised to their great dismay that the foundations on which mathematics was built and on which it had operated for millennia could no longer be guaranteed. The entire edifice of mathematics had to be rethought from scratch. One of the foundational theories of mathematics is algebraic geometry. And it was precisely in this field that Noether reduced the number of arguments and derivations that could only be arrived at in an arbitrary manner through continued abstraction until she had arrived at the underlying foundations of mathematics. This work can be compared to a continual questioning of the kind: "Why is that so?" It is only when we have asked it often enough that we arrive at the source. Emmy Noether's peerless capacity for abstraction has contributed significantly to the fact that mathematics today is internally consistent under certain assumptions.

In the early years, Noether had received more head-shaking than approval, but gradually the significance of her abstractions was recognised. In 1932, a few years before her death, she received what was then the highest award for mathematicians: the Ackermann–Teubner Memorial Prize (the Fields Medal has only existed since 1936). Deservedly, Emmy Noether is also called the "mother of modern algebra," although the title "mother of modern mathematics" would also be justified. Emmy Noether's ideas and her contributions to mathematics were so significant that her name became used as an adjective for many mathematical structures, such as "Noetherian rings," "Noetherian groups," and "Noetherian modules"—a similar accolade was also given, for example, to Newton with his Newtonian physics and Einstein with his theory of relativity. But, while simple popular accounts of Einstein's theory of relativity eventually became available to the general public, Noether's field of expertise was so far removed from any real world considerations that the public hardly ever became aware of her contribution, either during her lifetime or even today. Only within the small circle of mathematicians who work with the abstract theory-laden field of algebra is she still highly revered today.

The situation here is similar to what happened with the insights that Emmy Noether brought to physics. Throughout her life, she received only very limited recognition for tidying up the loose ends in the general theory of relativity.

- In his work on the special properties of GR with regard to conserved quantities, Felix Klein thanked Emmy Noether in a footnote, but accepted behaviour today would have been to include her as a co-author.
- The eminent mathematician Hermann Weyl, Hilbert's successor at Göttingen University, saw no need to explicitly address Noether's contribution in his book "Raum, Zeit, Materie—Vorlesungen zur allgemeinen Relativitätstheorie" (Space, Time, Matter—Lectures on General Relativity), published in 1918, or indeed in later revised editions.
- In Wolfgang Pauli's influential article on the theory of relativity, which appeared in the "Encyklopädie der Mathematischen Wissenschaften" of 1921, there is also no reference to Noether's work.
- Even Albert Einstein, who had praised her so highly in 1918, never cited Noether.

For eight years, Emmy Noether worked unpaid in Göttingen, just as she had done in Erlangen. In 1921, when she was 39 years old, her father died. Until then, he had supported her financially, and the salary she was finally awarded from 1923 was so low that she continued to live in precarious circumstances. Although suitable professorial chairs were up for grabs on several occasions, she continued to be denied a full professorship at a German university. She was not even allowed to keep her poorly remunerated position in Göttingen, because, when the National Socialists came to power in 1933, people of Jewish origin were no longer accepted in the civil service. Some of Göttingen's world-famous mathematics professors fell victim to the "Gesetz zur Wiederherstellung des Berufsbeamtentums" (Law for the Restoration of the Civil Service), which was passed in no time at all. Although Emmy Noether did not belong to the circle of established employees and this law was not to be applied to her, she too was forced out of her position. A stay in the USSR, where she had lectured as a visiting professor in 1928/1929 and strongly influenced the later world-famous mathematician Andrey Kolmogorov, made the National Socialists suspect her of being a communist. Although Emmy Noether had been a member of the social democrate party (SPD) until 1924 and a convinced pacifist, there was only *one* convinced communist in her family: her brother Fritz. He emigrated to Tomsk, Russia, in 1934, where he received a professorship in mathematics. However, in one of Stalin's purges, he was arrested in 1937 and executed in 1941.

Emmy Noether left Germany for America, where she taught at Bryn Mawr Women's College in Pennsylvania. She also frequented nearby Princeton, where she once again met up with some of the most famous scientists of the German-speaking world, including Albert Einstein and Hermann Weyl. Less than two years later, she died unexpectedly after cancer surgery. She was 53 years old.

Shortly after her death, Albert Einstein wrote in the *New York Times*:

> Noether was the most important creative mathematical genius so far produced since the beginning of higher education for women.[5]

This is a heartfelt and well-meaning compliment from Einstein, but it draws attention to Noether's gender rather than acknowledging that she clearly stood out among *all* mathematicians, not only female ones. Other colleagues also expressed praise, but could not refrain from simultaneously making remarks about her rotund figure and her lack of relationships with men. Only Hermann Weyl, one of the greatest mathematicians of the time, managed to provide an insightful and dignified obituary. He was personally present and spoke during the small funeral service in Bryn Mawr:

> I was ashamed to occupy such a privileged position next to her, whom I knew to be superior to me in many ways as a mathematician. (...) You were a great woman mathematician, I have no hesitation in calling you the greatest that history has known. Algebra has been transformed through your work.[6]

---

[5] From Einstein's obituary of Emmy Noether, which appeared in the *New York Times* on 4 May 1935 and was originally written in German.

[6] Memorial address Weyl gave to Emmy Noether at Goodhart Hall in Bryn Mawr on 26 April 1935.

12

Grete Hermann (1901–1984): Philosopher of Quantum Physics

At first glance the pioneers of quantum physics and relativity may seem to be an all-male club: Max Planck, Albert Einstein, Niels Bohr, Erwin Schrödinger, Werner Heisenberg, Wolfgang Pauli, Max Born, Paul Dirac, Enrico Fermi, and later Richard Feynman and Murray Gell-Mann are perhaps the best known. But the names Emmy Noether, Marie Curie, and Lise Meitner remind us that women also made significant contributions to these new fields of knowledge. And in addition to these three outstanding female scientists, there is a fourth woman whose achievements have only recently been rediscovered: Grete Hermann.

Hermann had learned mathematical thinking to the highest level of abstraction and rigour from Emmy Noether, and she also received her doctorate from her. Due to the extraordinarily good reputation of her doctoral supervisor, Grete Hermann could have ventured into a scientific career, but she did not only excel in mathematics. Her heart belonged to philosophy. After her doctorate in mathematics, she first turned her attention to quantum physics, because here there were significant overlaps with her favourite subject due to the many seemingly insoluble contradictions. For Hermann, the answers to the complex questions that have turned the known world upside-down since the discovery of quantum physics lay not in mathematics and physics, but in philosophy alone.

This attitude was not as far-fetched as it might seem. Until the 1930s, quantum physics was dominated by German scientists and the historical connection between physics and philosophy was strong. Great physicists like

L. Jaeger, *Women of Genius in Science*, https://doi.org/10.1007/978-3-031-23926-7_12

Einstein, Heisenberg, and von Weizsäcker were quite naturally philosophically minded. Nevertheless, Grete Hermann was the first person able to grasp the facts which characterise quantum physics, but which seem contradictory to our everyday conceptions, in terms of the humanities. Unfortunately, her far-reaching considerations long remained almost unnoticed. Today, many physicists suspect that the historical development of quantum mechanics would have been quite different if her insights had been taken into account during her lifetime.

* * *

Grete Hermann first saw the light of day in the same year as the physics genius Werner Heisenberg: she was born on 2 March 1901 into a strictly religious merchant family in Bremen as the third of seven children. Like her contemporaries Emmy Noether and Lise Meitner, she chose for lack of alternatives to train as a teacher after her Abitur, which women were still only allowed to take in exceptional cases. At the age of 20, i.e., within a very short time, she acquired the teaching qualification required for elementary and secondary schools, but never practised the profession she had trained for. Instead, she studied mathematics, physics, and philosophy in Göttingen and Freiburg. Only four years later, Grete Hermann received her doctorate in mathematics from her doctoral mother Emmy Noether.[1] However, she must have felt unsure about spending her life in the prevailing misogynistic academic environment, because in the same year she also took the examination for the teaching profession at higher schools. And in the end, she decided on a third path: in 1926, she became an assistant to the Göttingen philosopher Leonard Nelson, whose field of research was Kant's philosophy.

In the following years, Hermann also discovered the fascinating world of quantum physics. Here, too, she proved her great talent. Soon she was corresponding with great physicists such as Carl Friedrich von Weizsäcker, Werner Heisenberg, and Niels Bohr about the completely new connections that were becoming apparent in the world of the smallest particles. In the quantum world:

- In many cases, objects cannot be attributed unambiguous values of physical quantities. In 1925, Heisenberg showed with his uncertainty principle that it is impossible, for example, to precisely determine the position and momentum of an elementary particle at the same time.

---

[1] Grete Hermann, The question of finitely many steps in the theory of polynomial ideals, *Mathematische Annalen*, 95 (1926) pp. 736–788.

- Processes are not described using categories such as position, direction, and duration, but with statistical probabilities. This also eliminates the direct causal relationships known from Newtonian physics.
- There is no objective reality. Things and processes only become real when they are measured.
- Objects can influence each other even if they are connected neither by matter (like a hammer hitting a nail) nor by forces (iron filings are made to move by a magnetic field).

In the so-called Copenhagen interpretation, the Danish physicist Niels Bohr and his colleagues explained that there are two worlds with completely different laws: on the one hand, the world in which we humans find our way around with the help of Newtonian physics and which can be calculated according to its formulae, and on the other hand, the quantum world in which there are neither causal relationships nor objective reality.

Albert Einstein and Erwin Schrödinger were bitter opponents of this interpretation, because they were convinced that there could not be two worlds, but only one. Einstein assumed that there must be certain factors that had not yet been taken into account in the newly found laws of the quantum world. If they were found and the quantum laws were thereby supplemented, one would see that causality, objectivity, and reality also reign among elementary particles and atoms. He called these factors "hidden variables" and searched for them until his death.

* * *

The question of whether hidden variables exist and, if they do, whether or not they are detectable at all, divided physicists into two camps. Einstein's discussion with his Copenhagen colleague Niels Bohr was one of the most important philosophical disputes of the twentieth century. For years they tried to convince each other with ever new arguments. In 1932, the battle suddenly seemed to be decided: John von Neumann, who at 28 years old was already considered one of the greatest mathematicians of the day, published his epoch-making book "Mathematical Foundations of Quantum Mechanics," in which he proved that the mathematical structure of quantum theory definitely excludes hidden variables. Even though Einstein never accepted the consequences of this proof, the Copenhagen interpretation now prevailed as the common explanation of quantum theory. For many physicists, it is still valid today.

But there was a big problem with the proof by the great John von Neumann: it was wrong. In the 1930s, however, it did not occur to anyone to check the great master's derivation in detail or to try to contradict him. Only one person recognised early on the error in von Neumann's mathematical derivation: Grete Hermann.

Hermann was particularly interested in the philosophical question of whether the principle of causality could be upheld for atomic processes or whether, at this end of the size scale, only statistical probabilities could actually describe the processes. In her paper "Die naturphilosophischen Grundlagen der Quantenmechanik" (The Natural Philosophical Foundations of Quantum Mechanics), published in 1935[2] and previously circulated at least among the physicists in Leipzig where she was working with Heisenberg, Grete Hermann followed the philosophy of Immanuel Kant and came to a conclusion similar to Einstein's: quantum theory could not exclude the existence of as yet unknown variables. This result was in stark contradiction to von Neumann's proof. Had Grete Hermann not noticed that the matter had already been decided for two or three years in the eyes of the majority of the scientific community? Did she dare to openly oppose the world's greatest mathematician?

The matter was even more humiliating for John von Neumann. Grete Hermann dealt with the fundamental error in the mathematical luminary's assumption in just one short paragraph of her paper. Indeed, she brushed the proof aside in a few sentences[3]:

> For the expected value function E(R) thus defined for an ensemble of physical systems, which attributes a number to each physical quantity, von Neumann assumes that E(R + S) = E(R) + E(S). In words, the expected value of a sum of physical quantities is equal to the sum of the expected values of the two quantities. Neumann's proof stands or falls with this assumption.[4]

Physical expectation functions are the mean values of physical quantities, in this case the quantities R and S (for example, position and momentum). They are used when specific values are not known and one must use probability distributions of the values. The equation E(R + S) = E(R) + E(S) is universally valid in classical physics, but its validity in quantum physics was only an assumption. With her incorruptible eye, Grete Hermann immediately recognised this fundamental lapse and also knew that von Neumann's

---

[2] Grete Hermann, *Die naturphilosophischen Grundlagen der Quantenmechanik*, §7, Abhandlungen der Fries'schen Schule (ASFNF), Vol. 6, issue 2 (1935), p. 76.

[3] The author has adapted the notation in the quote for better readability. The expectation function is denoted here by E(R), rather than (Rφ,φ).

[4] Grete Hermann, *Die naturphilosophischen Grundlagen der Quantenmechanik*, §7: "Der Zirkel in Neumanns Beweis", Abhandlungen der Fries'schen Schule (ASFNF) 6, Heft 2 (1935), p. 99.

proof automatically lost its validity. In fact, we know today that the assumption is wrong: in quantum theory, there are many situations in which this equation is *not* valid.

As a trained philosopher, Grete Hermann was of course aware that, from her conclusion that hidden variables are *possible*, it cannot be concluded that they actually exist. The fact that von Neumann's formulae could not be applied to the quantum world was rather proof for her that the question of whether hidden variables exist or not could not be mathematically derived and proven. As it turned out thirty years later, she was right in this view.

* * *

What made the Copenhagen interpretation so interesting for the philosophical mathematician and mathematical philosopher Grete Hermann? This attempt to explain the quantum world had several fundamental problems:

- The two worlds of the Copenhagen interpretation must meet somewhere, on some scale of magnitude. But where exactly do the laws of classical physics merge into the completely different laws of quantum physics? Do they change abruptly above a certain size of the observed systems? Can there be areas where they overlap?
- The objects we know from our everyday experience can be understood in terms of Newtonian physics. But in every case, they are also composed of the smallest components of matter, and these obey quantum laws. So how does that fit together?
- In the world of Newtonian physics, measurements influence observations; for example, if the temperature of water is determined using a thermometer, the temperature of the measuring device will change the temperature of the object being measured by a small amount. In the quantum world, the measurement not only influences what is measured, it actually makes it real. It is only through measurement that quantum objects can be assigned properties with certainty.
- When a measuring device interacts with the object to be measured, the separation between the observing subject and the observed object is abolished. At the same time, however, the measuring device, which belongs to the Newtonian world, has objective properties, i.e., properties that are independent of the measured object. While Werner Heisenberg, for example, did not want to acknowledge this clear contradiction, Grete Hermann was also more rigorous in her thinking on this point.

The first point in particular caused discussion in quantum physics. The supporters of the Copenhagen interpretation assumed a sharp division called the Heisenberg cut, which separated the two worlds from each other. But even Bohr could only give an unsatisfactory answer to the question of what exactly the difference was between the two worlds: above a certain magnitude, the laws of the quantum world "somehow" merged into the laws of classical physics and vice versa. It remained completely open where exactly the Heisenberg cut should be on the scale of magnitude:

- Max Born and Wolfgang Pauli assumed it to be on the size scale directly above quantum objects such as protons and electrons. According to this view, even macromolecules would obey Newton's laws.
- Niels Bohr assumed that everything below the measuring apparatus obeys quantum laws.
- John von Neumann thought that the cut was somewhere between the measuring apparatus and the observer.
- Werner Heisenberg speculated that the dividing line could be chosen arbitrarily, depending on the context of the experiment. He wrote:

The dividing line between the system to be observed and the measuring instrument is directly defined by the nature of the problem, but for obvious reasons cannot involve any discontinuity of the physical process. For this reason, there must be complete freedom in the choice of the position of the dividing line within the limits.[5]

The basis for these assessments was "Schrödinger's cat", Erwin Schrödinger's famous thought experiment from 1935:

A cat is locked in a steel chamber together with the following infernal machine (which must be secured against any direct interference by the cat): in the tube of a Geiger counting there is a tiny amount of a radioactive substance, so little that in the course of an hour perhaps one of the atoms will decay, but just as probably none; if this happens, the counting tube responds and activates a small hammer via a relay, which smashes a small flask containing prussic acid. If one has left this whole system to itself for an hour, and if in the meantime no atom has decayed, one will say to oneself that the cat is still alive. The first atomic decay would have poisoned it. The psi-function of the whole system

---

[5] Werner Heisenberg, *Wandlungen der Grundlagen der exakten Naturwissenschaft in jüngster Zeit*, lecture to the Society of German Natural Scientists and Physicians, Hanover, 17 September 1934, *Angewandte Chemie* 47 (1934).

would express this in such a way that the living and the dead cat are mixed or smeared out in equal parts. The typical thing about such cases is that an indeterminacy originally confined to the atomic realm turns into macroscopic indeterminacy, which can then be decided by direct observation. This prevents us from so naively accepting a 'blurred model' as a reflection of reality.[6]

With this thought experiment, Schrödinger had shown that the laws of quantum physics can also creep into the size scale familiar to humans and thus into our everyday experiences. For in the cat experiment, we cannot separate the quantum world, in which the decision whether an atom decays or not is not simply "yes" or "no" but takes on a statistical value, and our own world of experience, in which a cat is definitely either alive or dead:

- If we consider the system "atomic decay–cat" according to the rules of the quantum world, it is only when we "measure", i.e., open the box and look, that we decide whether the cat has died or is alive. Before that, it is in an intermediate state between "alive" and "dead."
- If we consider the system "atomic decay–cat" according to Newtonian thinking, it should be possible to calculate unambiguously whether or not the glass flask will be shattered.

Neither variant makes sense. It was thus obvious that an arbitrary separation between the two worlds, as Bohr had proposed, could not be a solution. It was simply not self-consistent.

* * *

Surprisingly, Grete Hermann's refutation of von Neumann's proof had no consequences. Her realisation that the hidden variables were not yet off the table due to the defect she had found could have given new impetus to the discussion about the foundations of quantum physics. But perhaps physicists were quite happy not to have to continue the debate. It lost steam and finally died down completely.

There was another reason why the discussion about the basic principles of quantum physics ended after 1935. In Germany, a notion of "Aryan physics" was introduced; fundamental research and related philosophical considerations were no longer desirable. The focus was now on the—preferably military—application of the findings from physics. In addition, the previously unrestricted exchange between scientists of different nationalities

---

[6] Erwin Schrödinger, The Present Situation in Quantum Mechanics, *Naturwissenschaften* 23 (1935).

quickly dried up and even became almost impossible from 1939 with the beginning of the Second World War.

For the next thirty years, neither mathematicians nor theoretical physicists challenged von Neumann's proof. It was not until the 1960s that the Northern Ireland physicist John Bell realised that the possible existence of hidden physical variables in the quantum world had not been excluded after all. The corresponding paper was published in 1966.[7] His mathematical statements were almost identical to those of Grete Hermann. Perhaps he had come across her work during his research. But while Hermann had still been sure that the question could only be solved in the field of philosophy, Bell presented his famous inequality, now named after him, in his paper. It is a clearly defined, measurable criterion that would decide the question of hidden variables.

- If Bell's inequality is satisfied, there can be hidden variables in quantum theory.
- If Bell's inequality is violated, this proves that there can be no hidden variables.

The key thing about this inequality is that it can only be checked *experimentally*. So, Grete Hermann had been right about the inadequacy of von Neumann's proof and also about her assertion that the question of hidden variables could not be solved within theoretical physics. But she was wrong in assuming that philosophy alone could find the answer.

In the end, it was an *experiment* and neither *theoretical reasoning* nor *philosophy* that finally decided the battle over the interpretation of quantum theory. In 1982, the Frenchman Alain Aspect succeeded in testing Bell's inequality. The result: it was violated in the experiment.[8] This proved beyond doubt that there are no hidden variables in the quantum world. And it was also clear that in the quantum world there is indeed no reality independent of measurements, no objects independent of each other, and also no causality. This world is therefore fundamentally different from the world in which we humans find our way around. For the experiment, which first demonstrated this property of quantum matter, Alain Aspect received the 2022 Nobel Prize

---

[7] John Bell, On the Problem of Hidden Variables in Quantum Mechanics, Stanford Linear Accelerator Center (1964); published in: *Review of Modern Physics*, Vol. 38, No. 3 (July 1966), pp. 447–452; www.psiquadrat.de/downloads/bell66.pdf

[8] To be more precise: Only local hidden variables are ruled out by experiment. Nonlocal ones are theoretically possible.

in Physics. If Grete Hermann were still alive today, would she have received it with him? She would surely have deserved it.

* * *

For more than thirty years, our fundamental understanding of quantum physics was not significantly developed, but on the technological side, great successes were achieved, such as nuclear fission, the generation of laser radiation, and the miniaturisation of computer chips. Bell's work, however, brought fundamental research back into focus and people began to look again at the fundamental properties of quantum systems. This return to the scientific foundations enabled the discovery of many more properties of quantum particles, and in turn a whole new range of possible applications.

In particular, it is only recently that the existence of entangled particles has come under the spotlight, although they were predicted in the early years of quantum physics. Entanglement means that systems of two or more particles can no longer be described as a combination of independent one-particle states. They can only be specified by a common state. For physicists, this means that they cannot describe the individual particles of a system with separate equations, so-called wave functions, but must form a *single* wave function for all the particles involved. In principle, all the elementary particles in the entire universe would have to be taken into account in this wave function. In 1935, shortly before the discussion about the foundations of quantum physics broke off, Schrödinger had written:

> This property [entanglement] is not one, but the property of quantum mechanics, the one in which the whole deviation from classical thinking manifests itself, leading today to very exciting possibilities of entirely new quantum technologies![9]

With a thirty-year delay, the experimental and theoretical study of entangled particles from the 1960s onwards opened the door to the technologies mentioned by Schrödinger. Among others, Richard Feynman predicted the possibility of computers based on entangled particles at the first *Physics and Computation Conference* held at MIT in 1981.[10] Today, quantum computers are about to make a leap into some real applications.

---

[9] Erwin Schrödinger, Discussion of probability relations between separate systems. *Proceedings of the Cambridge Physical Society*, 31, 55 (1935).

[10] Richard Feynman, Simulating physics with computers, *International Journal of Theoretical Physics* 21(1982), pp. 467–488.

It is regrettable that Grete Hermann's considerations were not taken note of for so long. If fundamental research had been continued, these successes would probably have come much sooner, and perhaps we would already be working with private quantum computers. The question is: Why was Grete Hermann's work ignored, while John Bell's research over thirty years later immediately ushered in a new upswing in fundamental quantum physics research? There are a few explanations for this:

- Grete Hermann mentioned her criticism of the general validity of von Neumann's proof only very briefly. She was less concerned with mathematics than with the philosophical implications of quantum physics. In later publications, she no longer mentioned the problem she had identified.
- She published her work with a rather insignificant publisher. Although the article also appeared shortly afterwards in a much more renowned scientific journal, it was an abridged version that did not include the refutation of the proof.[11]
- After the Second World War, the language of physics was no longer German, but English. New generations of physicists had no access to Hermann's work, which was never translated into English.

These points provide an explanation for why Hermann's article remained forgotten *after the war*. But at the time of publication, the pioneers of quantum theory, almost all of whom were German speakers, could well have addressed Grete Hermann's criticism of John von Neumann's proof. Most of them were in close contact with her at the time. It therefore remains open: Why did Bohr, Heisenberg, and von Weizsäcker, but also Schrödinger and Einstein not take note of Grete Hermann's remarks on the clear refutation of von Neumann's proof? The fact that she was a woman does not seem to have had any influence. Grete Hermann was taken seriously and appreciated by her male colleagues:

- Werner Heisenberg devoted an entire chapter of his famous autobiography "The Part and the Whole" to the philosophical discussion that he and other well-known scientists such as Carl Friedrich von Weizsäcker had had with Grete Hermann.[12] Among other things, he wrote that the "young philosopher" had "imparted important insights to him and his colleague Friedrich von Weizsäcker".

---

[11] Grete Hermann, Die naturphilosophischen Grundlagen der Quantenmechanik, *Die Naturwissenschaften*, 23, 42 (1935) pp. 718–721.

[12] Werner Heisenberg, *The Part and the Whole*, R. Piper & Co. Verlag (1969).

- In 1934, Grete Hermann was a participant in the seminar led by Werner Heisenberg in Leipzig, which lasted several months and in which numerous other renowned physicists discussed quantum physics. Grete Hermann's publication, in which the refutation of von Neumann's proof can also be found, emerged from this seminar.
- In 1936 she received the prize of the Avenarius Foundation in Leipzig for her work "What consequences do quantum theory and field theory in modern physics have for the theory of knowledge?"[13]
- Her intensive correspondence over a long period of time with Carl Friedrich von Weizsäcker, Werner Heisenberg, and other physicists on philosophical topics is also well-established.[14]

As a scientist, it is thus clear that Grete Hermann had gained access to the "men's club" of physics. It would have been quite possible for her to make herself heard. That she did not do so was initially due to the fact that philosophy was more important to her than fundamental research in physics. Soon this preoccupation also paled before other topics that would later dominate her life: politics and pedagogy. Hermann drew ethical consequences from Kant's philosophy, which she applied in direct and practical ways: she actively fought against the National Socialists, who had come to power in Germany in 1933. In 1936 the danger became too great for her and she emigrated to England. In London, she entered into a marriage of convenience with a man named Edward Henry in early 1938 in order to obtain British citizenship. Grete Herrmann was now called Grete Henry-Hermann. Shortly before she returned to Germany in 1946, she divorced him and took her old name again.

She first became a teacher in her hometown of Bremen and was soon appointed professor of philosophy and physics at the new University of Education, in whose establishment she had been a significant driving force. Here she was able to give a decisive impetus to the educational and ethical training of a future generation of teachers. She was also involved in the founding of the Education and Science Union (GEW).

She also remained active in politics. She joined the SPD and worked with Willi Eichler on the Godesberg Programme of 1959, a party programme that emphasised free-market values and ethical considerations. The reorganisation of social democratic politics can therefore be traced back to a large extent to Grete Hermann's ideas. She was highly respected for her socio-political

---

[13] Kay Herrmann (Ed.), *Grete Hermann: Philosophy—Mathematics—Quantum Mechanics*, Springer (2019).
[14] Ibid.

commitment. From 1961 to 1978, she chaired the Frankfurt Philosophical-Political Academy.

In April 1984, Grete Hermann died at the age of 83. Until the very end of her life, she had been actively involved with philosophical issues. However, her early contribution to the interpretation of quantum physics remained forgotten for a long time. The fact that John Bell likely referred to her, although he did not mention her when he brought fundamental research out of its slumber, has only been appreciated in recent years. If one could ask Grete Hermann whether this late recognition means anything to her, she might well answer that her other achievements seem more important to her. The pass for the goal that John Bell scored three decades later, she provided somnambulistically along the way.

13

Chien-Shiung Wu (1912–1997): Queen of Experimental Nuclear Research

So far, this book has presented women scholars who, without exception, belong to the European tradition of thought. From Hypatia in Alexandria, whose hometown was strongly influenced by Rome during her lifetime, to the female scholars from the United States, who from the second half of the twentieth century onwards generally overshadowed their European colleagues and find their place in the following chapters, the people I discuss belong, up until the 2000s, almost exclusively to the cultural sphere deriving from Greco-Roman antiquity. The Chinese Chien-Shiung Wu, who undoubtedly belongs in the ranks of outstanding female scientists, was discriminated against in two respects: as a woman and also as one of the few female researchers to come from a different culture.

For a long time, European and American dominance in the economic and scientific fields prevented people from other parts of the world from being perceived as researchers of equal standing. Yet this supremacy only emerged in the early seventeenth century, which is not so long ago. At that time, one of the most significant revolutions in human history was taking place in Europe: religious dogmas were being replaced by modern scientific thinking. The resulting technological progress made Europe and later the USA the undisputed economic and military world powers. Western-generated knowledge became the standard.

Philosophers, historians and sociologists agree that there is a connection between the success of scientific thinking and the emergence of open societies in which people can freely develop their creativity. It is no accident that

© The Author(s), under exclusive license to Springer Nature
Switzerland AG 2023
L. Jaeger, *Women of Genius in Science*,
https://doi.org/10.1007/978-3-031-23926-7_13

Newton's laws opened the way to the Age of Enlightenment. To this day, free societies are world leaders in research. Conversely, in the 1930s, when many German researchers turned their backs on what was rapidly becoming a totalitarian country and science was being enslaved to ideological dictates, Germany immediately lost its long-standing role as world leader.

Just as in the days of the empire and later under Mao, China is still a totalitarian society by any possible definition. The brief awakening when some Chinese philosophers and politicians tried to impose a Western-style democracy in the first third of the twentieth century ended quickly. The country's successes in the field of scientific research are correspondingly marginal. Since the introduction of the Nobel Prizes in 1901, 616 natural scientists have been awarded this highest prize in the sciences. Only one person among them is from China, was educated there, did research there, and had Chinese citizenship at the time the Nobel Prize was awarded: this was the physician Tu Youyou who, in 2015, received the Nobel Prize for her fight against malaria. All other Chinese-born Nobel Prize winners were no longer Chinese citizens at the time they made their prize-winning scientific achievements and owed at least a large part of their education to Western universities. The situation is similar in mathematics: not a single Chinese mathematician has received either of the two highest awards in the field, the Fields Medal or the Abel Prize.

The fact that China has been able to develop into a *technological* superpower is another matter entirely. The country has developed very successful industrial production and processing methods on the basis of freely accessible scientific knowledge gained from other countries and has made great efforts to gain a significant place in the world community. In many fields, from AI to biotechnology and genetic engineering, from quantum technology to photovoltaics and nuclear fusion, Chinese scientists are applying knowledge gained in the West.

* * *

Chien-Shiung Wu (or according to the newer spelling, Jiànxíong Wú) lived in China at a time when the free development of Chinese society was still conceivable. She was born on 31 May 1912 near Shanghai. Her mother Fan-Hua Fan was a teacher and, contrary to Confucian tradition, was convinced that sons and daughters should have the same educational opportunities, an attitude which testifies to the upheaval China was undergoing at that time. Chien-Shiung Wu's father, Zhong-Yi Wu, was also progressive. He was actively involved in Sun Yat-sen's revolution, which put an end to the Chinese Empire in 1911 and turned the country into a republic. He founded a school for girls where his daughter received her primary education. At the age of ten, the highly intelligent Chien-Shiung transferred to a secondary school for girls, where she was taught according to the Western model. Wu's wealthy family could easily have simply bought her a place at this school, but Chien-Shiung took a regular entrance examination and came ninth out of about ten thousand applicants. She learned English at this school, where American teachers also taught. In 1929, she finished her school career at the top of her class and in 1930 began studying mathematics at the National Central University in Nanking, about 270 km west of Shanghai. However, she soon switched to physics.

After completing her physics degree in 1934, Chien-Shiung Wu moved to Zhejiang University, about 180 kms southwest of Shanghai, where she worked for two years as an assistant in the field of X-ray structural analysis. Her professor Gu Jing-Wie had completed her doctorate at the University of Michigan and encouraged Wu to take the same step. The application from the 24-year-old Wu was accepted at Michigan. The plan was to stay in the USA only temporarily, but world events would prevent Chien-Shiung Wu from returning to her crisis-ridden homeland for many decades:

- In 1937, the Japanese occupied China, taking advantage of its abundant raw materials and brutally exploiting the population. The occupation did not end until 1945.
- During the Second World War, Chien-Shiung Wu's involvement in the Manhattan Project, which aimed to build the atomic bomb, made it impossible for her to return to China.
- The clashes between the communists led by Mao and regular troops, reminiscent of a civil war, which had begun in the 1930s, continued until 1949.
- After the Communists finally took power, Chien-Shiung Wu decided to stay in the USA forever. She became a US citizen in 1954.

Any visit home remained unthinkable for a long time. Even though her family survived the turmoil of those years, Wu never saw her parents and her two brothers again. Above all, the separation from her father Wu Zhong-Yi, to whom she was very close, must have caused her much pain.

* * *

In August 1936, Wu arrived in San Francisco by ship. Here she soon realised that, in contrast to China, which was very progressive in this respect, women were considered intellectually inferior to men. She also learned that

Michigan was particularly strict about gender segregation, so she chose to work at the more liberal University of California at Berkeley. Here she met Chia-Liu Yuan, who had adopted the Americanised first name Luke. He was a grandson of Yuan Shikai, a former mortal enemy of her beloved father. Yuan had been commissioned as military governor and prime minister under the last Chinese emperor to put down the uprising, but he had then sided with the revolutionaries and become China's first president. Later, his attempt to found a new dynasty as emperor failed. The daughter of the revolutionary and the grandson of the traditionalist fell in love and married six years later.

Through Yuan's mediation, the head of the physics department at UCLA offered Chien-Shiung Wu a place in the graduate programme, even though the academic year had already begun. Yuan also introduced her to Ernest Lawrence, who was director of the radiation lab and later became her PhD supervisor. While Wu was still working on her doctorate in his lab, Lawrence was awarded the Nobel Prize in Physics for his invention of the cyclotron particle accelerator.

Chien-Shiung Wu was now in a delicate situation. In her home country, the Japanese were wreaking havoc. If she failed one of the exams in her doctoral programme, her residence permit might be put in question. Under great pressure, she threw herself into her work and was usually still to be found in the lab at 4 am. She thus made correspondingly rapid progress. For example, she studied the transformation of the radioactive isotope xenon-137 into caesium-137 with the Italian physicist Emilio Segrè. They had produced the necessary starting material through the nuclear fission of uranium in the cyclotrons invented by Lawrence. The decay of xenon-137 is an example of the beta decay discovered by Enrico Fermi in 1933, in which new elements are created by the conversion of neutrons into protons, and electrons are emitted in the process as beta radiation.

Segrè quickly recognised Wu's brilliance and even compared her to Marie Curie, whom Wu deeply admired. Lawrence, too, was well aware of his collaborator's abilities, and described her as the most talented experimental physicist he had ever known.[1] In 1940, at the age of 18, Wu completed her dissertation, which consisted of two parts:

- In the first part, she dealt with the beta decay of the radioactive isotope phosphorus-32 and the effect of what is known as *bremsstrahlung*, electromagnetic radiation emitted by the resulting electrons.

---

[1] Chiang, Tsai-Chien, *Madame Chien-Shiung Wu: The First Lady of Physics Research*, World Scientific Publishing Company (2013).

- In the second part, she described her experiments with xenon-137, which she had carried out with Emilio Segrè.

The dissertation committee, including her official supervisor Lawrence and the later head of the US atomic bomb project, Robert Oppenheimer, were particularly impressed by the second part of the dissertation.

With Oppenheimer (affectionately called "Oppie" by Wu), one of the world's best and best-known physicists expressed unusual enthusiasm about Chien-Shiung Wu's abilities and called her *the* authority in the field of beta decay. And yet the press reported little about her scientific abilities, instead referring to quite different qualities. Hardly a reporter refrained from describing how pretty and exotic Wu looked. An article in the *Oakland Tribune*, for example, described her as a "petite Chinese girl" who "looks as if she might be an actress or an artist or a daughter of wealthy stock in search of occidental culture."[2] The condescension in this article is striking:

She was so passionate and excited whenever "China" and "democracy" were referred to.

* * *

Like her European colleagues, Wu had to endure the disdainful attitude of representatives of the opposite sex at the beginning of her career. In her case, there was an additional circumstance that challenged her position more than that of other women. Chinese men and women have a long history of being discriminated against in the US. The only federal law in the United States to date that has ever prevented an entire ethnic or national group from immigrating to the United States is the *Chinese Exclusion Act of 1882*, which prohibited the immigration of Chinese workers for many years. In 1943, this law was replaced by a quota system; only since 1965 has the immigration of Chinese people to the USA been possible without quotas.

During the time when Wu was building her scientific career in the USA, resentment against the Chinese was omnipresent. In addition, in December 1941, one year after Wu had earned her doctorate, Japanese planes attacked Pearl Harbor. This act of war caused the general distrust of Asian-looking people to grow even more. The fact that the Chinese had been suffering from the occupation of Japanese troops in their home country for years made no

---

[2] *Oakland Tribune* (today, the *East Bay Times*) of 26 April 1941: *Outstanding Research in Nuclear Bombardment by a Petite Chinese Lady.*

difference to this. The omnipresent racism at the time did not distinguish between people of Japanese and Chinese origin.

As it happened, despite her brilliance and reputation, Wu was unable to find a research position and had to make a living as a physics lecturer. First, she taught on the East Coast at Smith College, where she was only allowed to teach women. She was rather unhappy and later moved to the Princeton Institute, again "only" as a lecturer. There, as the first female faculty member of the physics department, she taught naval officers and met celebrities such as Albert Einstein.

In 1944, she finally received an offer to take up research again. At the SAM (Substitute Alloy Materials) laboratory at Columbia University in New York, Wu was involved in the construction of the atomic bomb. With her work on radiation detectors, she contributed to the enrichment of weapons-grade uranium-235. Like most of the others involved in the Manhattan Project, however, she could only guess at the overall aims of this work.

In September 1944, Wu volunteered to work on the Manhattan Project even more intensively. After some less important work, she was finally called in for a central task. During the construction of the first nuclear reactor prototype, an unexpected problem had been encountered: it was switching on and off at regular intervals. Nobel laureate Enrico Fermi and another specialist in nuclear fission, John Wheeler, suspected that Xe-135, a product of uranium fission with a half-life of 9.4 h, was the cause. The two scientists remembered Wu's doctoral thesis from 1940, the second part of which had dealt explicitly with a radioactive xenon isotope. They contacted Wu and learned that she had prepared a paper in *Physical Review* explaining why Xe-135 intercepted more neutrons than expected and thus interrupted the chain reaction for the continuous fission of uranium (it turned out that Xe-135 had an unexpectedly high neutron absorption cross-section, thus interrupting the neutron-driven fission of uranium). Since it was important that Wu's findings should not to be made available to the enemy Germany under any circumstances, her work was withdrawn shortly before going to press and the papers were deposited in a well-secured location. Like most of the physicists involved, she distanced herself from her involvement in the Manhattan Project after the power of the atomic bomb had been revealed by the destruction of Hiroshima and Nagasaki.

* * *

After the war ended, in August 1945, in addition to her part-time job at Princeton, Wu took up a position as a professor doing only research, with no teaching, at New York's Columbia University. She remained loyal to New

York University for the rest of her working life. When her only son Vincent, who later also became a physicist, was born in Princeton in 1947, a valued friend and colleague came to Wu's hospital to welcome her child: Albert Einstein.

In November 1949, Wu began to work on Einstein's EPR thought experiment. Einstein vehemently rejected the entanglement of quantum particles predicted by Schrödinger as "spooky action at a distance", and Wu tried to approach this subject experimentally. In fact, she succeeded in demonstrating the existence of entanglement using a pair of photons. Above all, however, Wu, now established as a physicist and pursued her interest in beta decay:

- In 1934, Enrico Fermi had published his theory of beta decay, according to which the neutral particle in the atomic nucleus known as the neutron can transform into a positively charged proton, emitting an electron (and an antineutrino).
- A 1937 experiment by Luis Walter Alvarez showed that, conversely, a neutron (plus a neutrino) can be created from a proton plus an electron.

Wu repeated Fermi's and Alvarez's experiments and in 1949 was able to show how the various reactions associated with beta decay worked in detail, thus completely demonstrating Fermi's theory.

Her most significant achievement, however, was an experiment in 1956, with which she disproved the previously widely accepted principle of parity conservation in physics. What is the concept of parity in physics? In our world of experience, the phenomena in two physical systems that are mirror images of each other will run identically. A spinning top, for example, obeys the same laws of classical physics regardless of whether it is turned to the left or to the right. For a long time, it was assumed that the principle of parity conservation also applies to processes that take place at the atomic size scale. Here, however, forces are at work that play no role in our world of experience: the strong and the weak nuclear forces. In 1956, the two young theoretical physicists Chen-Ning Yang and Tsung-Dao Lee discussed the possibility of parity violation for the first time. Specifically, the problem was called the Θ-τ conundrum (theta-tau conundrum): when one of the newly discovered K^+ mesons decayed, theoretical indications suggested that there were two different, very short-lived decay products:

- Sometimes, the K + meson decayed to a Θ^+ particle (theta particle),
- Sometimes, it decayed to a τ^+ particle (tau particle).

At first, it had been assumed that they were two different particles, because that was how the reflection symmetry was given. But a closer look revealed that the Θ-particle and the τ-particle were identical—and yet distinguishable, since they were not mirror images of each other. Yang and Lee suspected that the parity principle could not be universally valid, but was violated when the weak nuclear force was involved.

For the experimental demonstration of their idea, they needed Chien-Shiung Wu, the "First Lady of Physics." Wu carried out the essential part of the experiment that demonstrated parity violation using the beta decay of radioactive cobalt-60. Wu exposed the cobalt isotope to a strong electro-magnetic field so that its atomic nuclei were all aligned along the same axis. This experimental step required extreme cooling of the cobalt to 0.003 K, a temperature only a tiny fraction above absolute zero. For this, Wu had to travel to the headquarters of the National Bureau of Standards in Maryland, because only there was the necessary equipment available to carry out the experiment. Wu observed the behaviour of the particles emitted when the cooled cobalt nuclei decayed. Some shot out in the spin direction, others in exactly the opposite direction. According to parity conservation, the number of particles in the two groups should have been equal. However, Wu discovered that more particles shot out in the direction opposite to the spin. This proved that the law of parity can be violated in nature.

The fact that 36-year-old Yang and 31-year-old Lee were awarded the Nobel Prize in Physics in the same year shows the significance of this finding. Wu's substantial contribution, however, was not taken into account. In general, this decision by the Nobel Prize Committee was received with incomprehension. And the physicist Gell-Mann, who will be mentioned below, asked Wu:

How long did Yang and Lee pursue you to follow upon their work?[3]

Yang and Lee themselves also felt that Wu deserved the honour as much as they did. Yang later said that Wu had been the only person who understood the urgency and importance of what they were doing and then found a way to prove the parity violation.

* * *

[3] McGrayne, Sharon Bensch, *Nobel Prize Women in Science: Their Lives, Struggles, and Momentous Discoveries,* Carol Publishing Group, Secaucus, NJ (1993), p. 278.

Even though Wu was not awarded the Nobel Prize in Physics, her experimental proof of the violation of the law of parity finally made her an icon of experimental physics; her experiment is still remembered today as the Wu experiment. In the following years, Wu made further spectacular discoveries and received numerous awards for her contributions to nuclear physics:

- In 1952, she was appointed associate professor, becoming the first woman in the history of Columbia University to hold a tenured professorship in physics.
- In 1958, she became the first woman to receive an honorary doctorate from Princeton, and in the same year she was promoted to full professor at Columbia University.
- In 1963, she was awarded the Comstock Prize of the *National Academy of Sciences*, which is awarded every five years.
- In 1973, she was appointed the first president of the *American Physical Society.*
- In 1976, US President Gerald Ford awarded her the National Medal of Science.
- In 1978, she received the first Wolf Prize in Physics from the *Wolf Foundation* in Israel for the Wu experiment—this prize is now commonly regarded as a consolation prize for scientists who have not won a Nobel Prize.

After providing an improved version of Fermi's 1934 model of beta decay in 1949, she went even further in December 1962. Through an ingenious experiment, she demonstrated an even more universally valid form. This was an important building block for two other giants of nuclear physics: Richard Feynman and Murray Gell-Mann.

Physicists have long dreamt of unifying all physical forces into a single force. Currently, a distinction is made between four forces: the electromagnetic force, the gravitational force, and the strong and weak nuclear forces. All other forces, such as the friction force and the buoyancy force, can be traced back to one of these four forces. Feynman and Gell-Mann worked to unite the electromagnetic force with the weak nuclear force in a standard theory of particle physics (this was achieved in the late 1960s by the physicists Steven Weinberg, Sheldon Glashow, and Abdus Salam). Wu's 1956 proof that parity is not always conserved had already taken them a big step further. Chien-Shiung Wu's new version of beta decay from 1962 provided yet another important piece of the puzzle for Feynman and Gell-Mann's standard theory of particle physics, which was still under construction.

With her impressive ideas and abilities, Wu was considered by experimental physicists to be one of the best in her field. Herwig Schopper, Director General of CERN, once recounted the *bon mot* circulating in physics circles: "If the experiment was carried out by Wu, it must be right."[4] Experiments she conducted with students and staff in the 1960s and 1970s further substantiated this reputation:

- This included later investigations into double beta decay, where she recorded the properties of atoms in which muons take the place of electrons. These experiments took place half a mile under Lake Erie near Cleveland, Ohio, in a salt mine.
- The broad spectrum of her interests is demonstrated by her later contribution to research into sickle cell anaemia, which is caused by the deformation of haemoglobin molecules.
- Wu also worked on Bell's theorem, which promised a universal interpretation of the particle entanglements in quantum theory. This was one of the few cases where other physicists succeeded in doing something she had not been able to achieve: in the early 1980s, the Frenchman Alain Aspect succeeded experimentally in demonstrating the existence of the particle entanglements in quantum theory—for which he received, as mentioned above, the Nobel Prize in Physics in 2022.

Just like her great role model Marie Curie, Chien-Shiung Wu also managed, through great talent and tireless work, not to be marginalised in the scientific community like so many other women. Quite the opposite: she consistently won the respect of her colleagues. Even after Wu retired in 1981, many physicists continued to ask her for advice on how to demonstrate their theories in practice. When physicists were asked to name the most important female representatives of their field, there were essentially only three names among the answers: Marie Curie, Lise Meitner—and Chien-Shiung Wu.

* * *

One facet of Chien-Shiung Wu's personality has not yet been mentioned: her political commitment. As the daughter of a revolutionary, she experienced a childhood shaped by politics. Besides her father, another person exerted great influence on her political thinking: the eminent philosopher and liberal politician Hu Shi. Like all students in China, Chien-Shiung Wu was obliged

[4] Chiang, Tsai-Chien, *Madame Chien-Shiung Wu: The First Lady of Physics Research*, World Scientific Publishing Company (2013).

to spend a year teaching in a public school as part of her studies. Hu Shi was the headmaster at the school in Shanghai that Wu chose. Hu became a fatherly friend and supporter and the two kept in touch until his death in 1962. From 1938 to 1942, Hu was China's ambassador to the US and lived in America permanently from 1948 when the communists came to power.

While studying in Nanking, Chien-Shiung Wu became involved in student politics and was elected to a leading position. As she was one of the best students at the university, the authorities were lenient about her activities. Later in the USA, she also took an open and self-confident stand on socio-political issues. For example, in 1964, at a symposium at MIT in Cambridge, Massachusetts, she asked the auditorium:

> I wonder whether the tiny atoms and nuclei or the mathematical symbols or the DNA molecules have any preference for either masculine or feminine treatment.[5]

In particular, she was keen to see science subjects taught to both boys and girls. She campaigned for human rights issues and actively protested against the imprisonment of various dissidents as well as the Tiananmen Square massacre in June 1989. The emotional connection to her home country was never broken. But it was only in her later years that she was allowed to travel to China and Taiwan.

She died of a stroke on 16 February 1997 at the age of 85.

---

[5] Zuoyue Wang, *Wu Chien-Shiung. Dictionary of Scientific Biography. Vol. 25*, New York, Charles Scribner's Sons (1970–1980), p. 367.

14

Rosalind Franklin (1920–1958): The Woman Next to Watson and Crick

Historians do not classify centuries according to the dates of the calendar, but according to historically significant developments. When they speak of the nineteenth century, they mean the long period whose beginning is marked by the French Revolution of 1789 and whose end coincides with the Russian Revolution of 1917 and the entry of the USA into the First World War in the same year. The twentieth century is much shorter in their eyes. It begins in 1917 and is considered to have ended with the year 1989, when communism collapsed in Eastern Europe. According to this way of counting, the chemist Rosalind Franklin, born on 25 July 1920, is the first of the important female scientists presented in this book to be born in the twentieth century.

Two names are commonly associated with the discovery of the molecular structure of deoxyribonucleic acid DNA in 1953: James Watson and Francis Crick. Rosalind Franklin, who had contributed significantly to the elucidation of the double helix structure of DNA, was long overlooked by historians. The fact that she was not considered for the Nobel Prize is doubly scandalous. For her colleague Maurice Wilkins, who together with Watson and Crick was awarded the most prestigious of all prizes, had stolen crucial documents from Franklin's laboratory and handed them over to the research duo Watson and Crick. Without these documents, the three scientists would only have been able to decode the DNA—if at all—much later.

In the early 1950s, Rosalind Franklin was one of the most important international specialists in the X-ray structural analysis of macromolecules. This method uses X-rays to detect the three-dimensional structure of molecules.

© The Author(s), under exclusive license to Springer Nature
Switzerland AG 2023
L. Jaeger, *Women of Genius in Science*,
https://doi.org/10.1007/978-3-031-23926-7_14

This is because, unlike visible light waves, for example, X-rays have extremely short wavelengths; indeed their typical wavelength corresponds roughly to the distance between atoms in a chemical compound. While light waves "rush" through the molecule, X-ray waves are diffracted by the electrons present in the molecule. The more electrons are in the way, the stronger the diffraction. Behind the substance sample, the resulting out-of-step waves form an interference pattern that can be made visible on a photographic plate. But what sounds quite simple in principle is actually a highly complex matter. For example, unwanted sources of interference have to be excluded with the utmost precision and complicated calculations have to be carried out in order to be able to draw conclusions about the molecule from the resulting diffraction patterns.

It is undisputed today that Rosalind Franklin's X-ray diagrams and the interpretations of the images she made were groundbreaking. The fact that she was denied recognition for this achievement had several reasons:

- King's College London, where Franklin was working at the time when DNA structure was being decoded, was even more committed to traditions than other institutes. Among these traditions, one was that women scientists were not considered full researchers.
- Rosalind Franklin worked very precisely. Differing views on the required accuracy of scientific work caused tensions in the King's College laboratory. In addition, there were serious conflicts within the team over hierarchical issues. Franklin kept her research results under lock and key to protect them until they could be reviewed according to scientific standards. Unfortunately, this precaution was in vain.
- The conflicts within the team at King's College London also meant that Franklin moved to another college and turned to other topics. Exactly at the time when her own manuscripts on DNA structure and those of Watson and Crick had been submitted but not yet published, she moved on from King's College. There was no one left to represent her interests.
- Rosalind Franklin died in 1958 at the age of only 37.

* * *

Rosalind Franklin was born on 25 July 1920 in the London district of Notting Hill, the daughter of one of England's richest families. Her parents made no distinction in the upbringing and educational opportunities they offered their two daughters and three sons. But they tacitly assumed that their daughters would later marry and start a family. They did not support Rosalind's wish to become a scientist.

From the very beginning, Rosalind Franklin showed exceptional academic performance. In her family, this was a cause of pride, but also some irritation. For example, one of her aunts wrote to her husband:

Rosalind is frighteningly clever—she spends all her time calculating for pleasure and always gets it right.

At the age of 11, she attended one of the few girls' schools where physics and chemistry were taught. She was the best in her class and won prizes every year. She decided to become a scientist at the age of 15. At this age she also decided that she would never marry and have children, even though she loved children very much. This point alone reveals her scrupulous consistency, which permeated her entire life. In 1938, she passed her school exams with six distinctions and received a university scholarship. In the same year, she began to study chemistry at Cambridge, finishing in 1941, again with distinction. During her studies she met two people who had a great influence on her:

- Bill Price, an expert in spectroscopy, introduced her to this research method.
- Adrienne Weill, a pupil of Marie Curie who had fled France to escape the Nazis, became a good friend and later helped her to get her first job.

Franklin was an over-achiever who left school and studies behind her on the fast track and with top grades. But she encountered serious problems on the way to her dissertation. Her doctoral supervisor in Cambridge was Ronald Norrish, who received a Nobel Prize in 1967 but was a heavy drinker when Franklin began her doctoral work in 1941 and had something against women in the lab. Again, Franklin displayed her usual unconditional consistency: she left Norrish's laboratory and moved to the *British Coal Utilisation Research Association* BCURA, where she did research on the properties of coal. She quickly became an expert and contributed, among other things, to increasing the efficiency of gas masks that were essential in the war effort. She received her doctorate from Cambridge in 1945. She also made a contribution to her country in other ways during the war: together with her cousin Irene, she volunteered at an early stage to go on patrols to report German air raids.

After the end of the Second World War, Franklin asked her friend Adrienne Weill what she could do as a "physical chemist who knows very little about physical chemistry but a lot about the holes in coal." In autumn 1946, Weill introduced her at a conference to the director of the French *Centre national de la recherche scientifique* (CNRS), the all-powerful network of scientific institutes in France. At the beginning of 1947, the CNRS gave Franklin a position at the national central laboratory for chemical services in Paris. The head of the laboratory was Jacques Mering, an expert in X-ray crystallography.

Crystalline structures had already been studied thousands of times and for many years using this method, but Mering and his team were working on using it to study non-crystalline substances as well—including living structures. To do this, the substances had to be converted into a crystalline state in a complex process. In Paris, Franklin would thus become a specialist in the X-ray structural analysis of crystallised macromolecules.

In 1951, Rosalind Franklin went back to England. John Randall, the head of the biophysics department at King's College London, offered her a three-year research fellowship. The original plan was for Franklin to study the diffraction of X-rays on proteins and lipids. But even before she arrived in England, Randall decided that there was a much more exciting field of research for her, namely investigating DNA.

* * *

For many decades, scientists had been trying to track down the genetic information in living beings. It was known that in the cells of plants and animals there must be some kind of structure, some chemical compound, which transfers the genetic material from mother cells to daughter cells. For the smallest, i.e., indivisible unit of this information, the word "gene" had been introduced in 1909. But nobody knew what a gene was. That the strange structures called chromosomes in the cell nucleus could be the carriers of genes was only one of many theories. For example, certain cell proteins were also suspected of passing on characteristics from one cell generation to the next.

As time went by, it gradually came to be suspected by more and more scientists that deoxyribonucleic acid, or DNA for short, must be the gene-carrying substance. However, this was anything but proven. In 1950, little was known about this giant molecule:

- The individual building blocks of DNA each consist of a phosphate residue, a sugar, and a basic nitrogen compound.
- Sugar and phosphate molecules alternate to form a chain, and the bases somehow attach to this chain.
- In the late 1940s, the American Nobel Prize winner Linus Pauling, at that time the leading luminary in biochemistry, had developed a three-dimensional model for certain proteins in which the building blocks of

the protein were arranged in a structure that resembled a spiral staircase. Pauling also considered such a helix for the DNA molecule.

- Some scientists suspected a double or even a triple helix.

In 1950, the Swiss chemist Rudolf Signer produced extremely pure DNA for the first time and sent samples to various scientists for further investigation. One of these was Maurice Wilkins, who worked at King's College. Together with his doctoral student Raymond Gosling, he succeeded in producing a diffraction image of the DNA sample with quite good quality using X-ray structural analysis. His supervisor John Randall was delighted that Rosalind Franklin would soon be joining the team, because he knew she would succeed in making even better images.

Rosalind Franklin arrived at King's College in January 1951, but she quickly discovered that female scientists were not accepted as equal colleagues. This was evident from the fact that, as a woman, she was denied access to the dining hall where all the other higher-ranking scientists could socialise with each other.

Unfortunately, her boss John Randall had not been clear in his communication. There is no other way to explain why Maurice Wilkins was firmly convinced that Franklin was coming to the college, not as his colleague, but as his assistant. He was anything but thrilled when Randall assignéd Gosling, a doctoral student, to assist Franklin in her research instead of him. Since Franklin was unwilling to give in and Randall didn't put his foot down either, Franklin and Wilkins clashed again and again. Soon the two of them would hardly exchange a word with each other.

Later, in his book "The Double Helix," James Watson described the situation like this:

> Maurice, a beginner in X-ray diffraction work, wanted some professional help and hoped that Rosy, a trained crystallographer, could speed up his research. Rosy, however, did not see the situation this way. She claimed that she had been given DNA for her own problem and would not think of herself as Maurice's assistant. (...) Clearly Rosy had to go or be put in her place.[1]

This passage is doubly interesting. First, the use of the name "Rosy" for Rosalind Franklin seems somewhat arrogant (even though Wilkins is also referred to on a first-name basis). Most importantly, even Watson, the man who never admitted how important Franklin's contribution had been to his

---

[1] James Watson, *The Double Helix*, A Personal Account of the Discovery of the Structure of DNA, Signet (1969), pp. 16 and 17.

success, unequivocally spells out who was the expert and who was the novice in King's College X-ray crystallography lab.

* * *

Rosalind Franklin had significantly more expertise in her field of research than her male colleagues. Since it had proved impossible for her to work in a team with Wilkins, she avoided him. For her experiments, which she conducted mainly with Gosling, she used a new fine-focus X-ray tube commissioned by Wilkins and also a microcamera, which she carefully refined and adjusted herself. One of her crucial innovations was the creation of a chamber in which she could control the humidity. With this experimental setup, she succeeded in taking far better pictures than those Wilkins had made with Gosling. Franklin spent much of her time solving certain technical problems to increase the resolution of the images.

In the spring of 1952, Gosling used Franklin's apparatus and procedure to take the photo with number 51, which contained crucial information for determining the structure of DNA. At first, the photo looked misleading. The DNA looked like a double helix! But there were several reasons why this would be problematic:

- A double helix would not be chemically stable.
- Such a complicated structure could not unwind itself in the process of cell division.
- It would be difficult for it to copy itself.

Nevertheless, Franklin trusted her photographs. In November 1952, she presented her latest X-ray images in London and also stated her conclusion that DNA must be a double helix. Sitting in the auditorium was 24-year-old James Watson. Together with Francis Crick, who was twelve years older, he had also been working flat out on deciphering the structure of DNA in Cambridge, only 100 kms away. They were taking a trial-and-error approach, using models made of spheres and rods to search for stable molecular variants. Dealing with the findings of other scientists was less their thing. For example, they overlooked a publication that had given DNA research an important boost: the chemist Erwin Chargaff had discovered that, in the DNA of every creature thus far studied, the amounts of the bases adenine and thymine as well as those of cytosine and guanine were the same. Only after some delay did Watson and Crick realise that Chargaff had long since solved one of their problems. A meeting of the three did not work out to the advantage of the

Crick–Watson research team—Chargaff considered the two to be ignorant beginners and "scientific clowns".

Rosalind Franklin comes to a similar conclusion when, shortly after her lecture, she was invited by Watson and Crick to Cambridge to look at the duo's latest model. Franklin realised at first glance that their proposal bore no relation to her research findings: Watson and Crick had built a triple helix with nitrogen bases pointing outwards. Watson had to admit that he didn't understand some of her lecture and—even more embarrassingly—he took down some of the details incorrectly. Indeed, he was thoroughly shown up that day. But this setback did nothing to deter the duo from continuing their research.

Franklin was upset that she had been asked to come to Cambridge simply because of this blunder and she was no longer willing to share her research results with Watson and Crick. She found their approach questionable, based as it was on building DNA models without sufficient data. Franklin relied solely on verifiable experimental data. Her doctoral student Gosling once quoted her:

> We're not going to speculate, we are going to wait, we are going to let the spots on this photograph [via the Patterson] tell us what the structure is.[2]

Franklin was working mainly on the tedious and labour-intensive task of getting to grips with what was known as the phase problem: with structures as small as the DNA building blocks, it matters which phase the X-ray waves are in when they hit the structure to be measured. By comparing a large number of photos with sometimes contradictory data, she was able to work out the phase shifts. By January 1953, Franklin had reconciled the data and was able to provide proof for her assumption that DNA was made up of a double helix and that the bases had to point *inwards*.

In the meantime, Watson and Crick continued to work with their construction kit. On 30 January 1953, they learnt through an indiscretion that the great Linus Pauling would publish an article presenting DNA as a double helix with the nitrogen bases on the *outside* of the molecule. Watson and Crick were shocked. Pauling had beaten them to it in the race to decode the DNA structure and would go down in history as the one who had first discovered it! They could see only one way out: they could bet on Pauling's

---

[2] Samuel Schindler, Model, Theory, and Evidence in the Discovery of the DNA Structure, *The British Journal for the Philosophy of Science*, published by: The University of Chicago Press, 59, (2008) p. 629; available at www.jstor.org/stable/40072305

solution not being the right one, while Franklin was right in assuming that the bases were on the inside of the helix.

Now this chapter in the history of science finally becomes a thriller. On the same day, Watson travelled to King's College with a preprint of Linus Pauling's article. He wanted to persuade Rosalind Franklin to work with them despite their differences. But Franklin refused. When Watson suggested that Franklin would not be able to interpret her own data correctly, the situation escalated. Watson left Franklin's lab angrily and bumped into Wilkins, who had often clashed with Franklin himself. Wilkins took sides with the Cambridge scientists and stole Franklin's research results from her office. Among them were the documents that proved that the bases were located on the inside of the helix, as Franklin had long suspected. So, Pauling had indeed made a mistake and Watson and Crick are back in the race.

Later, Watson and Crick denied having seen Franklin's crucial notes. Only fifteen years later, in his book "The Double Helix," did Watson admit to having obtained Franklin's data by dubious means and to having used it against her explicit will:

I knew more from her records than she thought.[3]

Within five weeks, thanks to Franklin's information, Watson and Crick completed the model on which their fame would subsequently be based. On 7 March 1953, the time had come: the model consisted of two helical strands of sugar-phosphate chains, in the interior of which the nitrogen bases were connected to each other in opposite directions. These relatively loose hydrogen bonds could break during the DNA reproduction process. The entire molecule would open like a zip, so that the nitrogen bases that were previously inside were now freely accessible and could be added up again by their counterparts. In this way, two double strands were created from one double strand, which could be distributed to the daughter cells. Watson and Crick's model was wonderfully simple and at the same time plausible—at last a structure had been found that combined all the information into a self-consistent mechanism!

Nevertheless, Franklin was still one step ahead of the research duo. After meticulously evaluating her recordings to make sure she had understood all the details, she wrote three papers:

---

[3] James Watson, *The Double Helix*, A Personal Account of the Discovery of the Structure of DNA, Signet (1969), p. 165.

- One reached the journal *Acta Crystallographica* in Copenhagen on 6 March 1953, the day before Crick and Watson completed their final model of DNA. So Franklin could not possibly have known about their results when she wrote her article.
- Another of Franklin's articles with Gosling as co-author went to the renowned journal *Nature* shortly afterwards.
- On 17 March 1953, she supplemented the *Nature* article with a third article in which she described the double helix structure in more detail. She was still ahead of Watson and Crick, who sent their manuscript to *Nature* barely two weeks later, at the end of March 1953.

The publications by Franklin and Gosling and by Watson and Crick appear in the same issue of *Nature* on 25 April 1953.[4] Crick and Watson admitted in a footnote to having been "stimulated by general knowledge of the unpublished paper by Franklin and Gosling." This was a gross understatement, for in fact their model relied heavily on Franklin's findings, thanks to the data pilfered by Wilkins.

As a result of an agreement between the journal *Nature* and the two laboratory directors in London and Cambridge, Franklin and Gosling's article was positioned behind Watson and Crick's in the *Nature* issue. This order reinforced the impression that Franklin's X-ray crystallography only served to support the work of Crick and Watson.

* * *

For a long time, Rosalind Franklin was portrayed as a dogged and aggressive lone wolf who was uncollegial in not wanting to share her research findings. But is it really true that she was not a team player? Or was it rather due to her male colleagues that she did not willingly provide the fruits of her achievement and thus confidently chose the only way not to be undervalued?

In mid-March 1953, her research fellowship expired and, as had been planned for some time, Rosalind Franklin moved from King's College, where her research was a daily struggle, to Birkbeck College, London University. In a letter to her friend Adrienne Weill in Paris, she described the move thus:

---

[4] James Watson, Francis Crick, A Structure for Deoxyribose Nucleic Acid, *Nature,* 171 (25 April 1953), pp. 737–738.

Rosalind Franklin, Raymond Gosling, Molecular Configuration in Sodium Thymonucleate, *Nature,* 171 (25 April 1953), pp. 740–741.

As for the lab, I'll be moving from a palace to the slums, but I'm sure I'll still find Birkbeck more pleasant.[5]

At Birkbeck College, she was once again able to show that she valued cooperation with other scientists. Here she met, among others, Aaron Klug, who had just completed his doctorate at Trinity College in Cambridge. She soon formed a successful collaboration with him and also a deep friendship. At Birkbeck, her achievements were also appreciated: at the end of 1954, the chair of physics offered Rosalind Franklin the opportunity to lead her own research group.

With the move from King's to Birkbeck, Franklin ended her work with DNA and turned to ribonucleic acid RNA, a molecule that was just as important for understanding life as DNA. Her research focused on the tobacco mosaic virus (TMV), which was well suited to X-ray crystallography because of its regular structure. Franklin showed that viral RNA is not a double helix like DNA, but a single helix.

In 1956, the now 36-year-old Franklin traveled with colleagues to the USA, where she visited scientists at the University of California at Berkeley. They suggested Franklin research another virus, the highly contagious virus responsible for polio, which had cost the lives of many children for many centuries and against which a vaccine had only come onto the market the year before. Franklin took up the suggestion and applied for corresponding research funds in 1957. The US health system actually granted for three years the largest sum Birkbeck had ever received. Together with her team, Franklin began to decode the RNA structure of the polio virus in a crystalline state. She also wanted to work with living rather than dead polio viruses. Due to the dangerous nature of the virus, serious reservations were expressed about this wish, but in July 1957 Franklin was given the green light.

The trip to the USA the year before, however, was not only under a good star. At that time, it became apparent that Franklin had contracted ovarian cancer, probably triggered by her continued exposure to X-rays. After just one year, her working capacity was so severely limited that Aaron Klug and his student John Finch have to complete her work.

Rosalind Franklin died in Chelsea on 16 April 1958 at the age of just 37. How much she was appreciated by her Birkbeck colleagues is shown by the article on the crystal structure of the polio virus published by Klug and Finch after her death. It is respectfully dedicated to her memory.

* * *

---

[5] Brenda Maddox, *Rosalind Franklin: The Dark Lady of DNA,* HarperCollins, London (2002).

Rosalind Franklin was an outstanding researcher who did her science in respect of all the rules of the art. She was a team player and recognised worldwide as an expert. One thing she was not, however, was vindictive. The breach of trust shown by the theft of her documents from her laboratory did not make her bitter. She willingly acknowledged Crick and Watson's soon worldfamous DNA model as their achievement. She later even became friends with Francis Crick and his wife. When Franklin later fell ill with cancer, she spent the time of her recovery with the Crick couple after several stays in hospital. James Watson, on the other hand, was far less empathic; in the 2000s, he made the headlines for unacceptable racist and anti-feminist statements that led to the revocation of some of his honorary degrees.

During Franklin's lifetime, the DNA structure was considered decoded, but not the exact processes by which this molecule translates genetic information into the synthesis of certain proteins. For example, it was not yet known that each of the 21 amino acids that make up proteins is encoded by a very specific base triplet. It was not until 1961 that Crick and his colleagues were able to prove this mechanism. If Franklin had still been alive, they would most likely have succeeded more quickly.

Rosalind Franklin also did not live to see James Watson, Francis Crick, and Maurice Wilkins receive the Nobel Prize for Medicine/Physiology in 1962 for decoding the structure of DNA. In the speeches given on the occasion, Watson, Crick, and Wilkins made no mention of their former colleague and her crucial recordings. Her contribution was forgotten; in the scientific literature, what Franklin had so vehemently opposed finally happened: she was referred to as Wilkins' assistant. It was only decades later that the dubious circumstances under which Wilkins had appropriated Rosalind Franklin's research results and under which it had been possible for Watson and Crick to complete their DNA model became known. Today it is undisputed that Franklin made a decisive contribution to the discovery of the DNA structure and that the theft of not yet published research results grossly violated the rules of good scientific practice.

Involved in this rectification was Aaron Klug, who had worked with Franklin at Birkbeck College. It was important to him to give Rosalind Franklin the credit she deserved. He examined Franklin's documents from the time when DNA was being decoded and came to the conclusion that she had indeed possessed all the necessary information about the DNA structure even before Watson and Crick. Unlike Watson and Crick, who, apart from the stolen information, had worked out their model on theoretical grounds alone, Franklin also had the proof in the form of experimental data.

Aaron Klug received the Nobel Prize in Chemistry in 1982 for his development of crystallographic electron microscopy and the elucidation of the structures of biologically important nucleic acid–protein complexes that was achieved with this method. This was precisely the field of research that Franklin had suggested to him almost a quarter of a century earlier and on which they had both worked for some time. Unlike Watson and Crick, however, he recalled Rosalind Franklin and the influence her research and advice had had on his life with great gratitude in his Nobel Prize speech.[6] He expressed the obvious point that, had his mentor been granted a longer life, she would not only have been awarded the Nobel Prize for the decoding of DNA, but would probably also have been a co-winner of his own Nobel Prize. Many other scientists also believe that Rosalind Franklin—like Marie Curie before her—deserved two Nobel Prizes and might well have received them had she not died so young.

---

[6] Available at www.nobelprize.org/uploads/2018/06/klug-lecture.pdf

15

Jane Goodall (*1934): The Great Lady of Primate Research

Jane Goodall is the first of the great female scientists featured in this book whom the reader can still see in action in the early 2020s. She was born on 3 April 1934, the same year that Marie Curie died, and at over 80 she is still going on lecture tours. For thirty years, the primatologist and anthropologist researched the behaviour of wild chimpanzees in Africa, and her findings have forced humanity to rethink its relationship to the animal world. Over forty films and numerous reports have made the name of Jane Goodall world famous.

Why chimpanzees in particular? This question has many answers. One of them is the stuffed toy monkey named Jubilee, which she received as a gift from her father, an engineer from a wealthy family, when she was a small child. The much-loved stuffed animal is still around today and sits in its place on a dresser in Jane Goodall's London flat. A few years later, another man inspired Goodall's dream of going to Africa and living among the apes: Tarzan.

> When I was 10 or 11, I found a second-hand book—we couldn't have afforded a new book—called Tarzan of the Apes, and I read it from cover to cover. Of course I fell in love with Tarzan. Of course he married the wrong Jane.[1]

She was also fascinated by the story of Doctor Dolittle, who could talk to animals. In addition to the fascination that the animal world exerted on her,

---

[1] Read more at www.britannica.com/explore/savingearth/the-right-jane.

© The Author(s), under exclusive license to Springer Nature
Switzerland AG 2023
L. Jaeger, *Women of Genius in Science*,
https://doi.org/10.1007/978-3-031-23926-7_15

there was her urge to explore, which she showed even as a small child. When she was four years old, the family went on holiday to her grandmother's farm. One day, little Jane could not be found for hours. Only that night could the family give the all-clear and send the police home. Jane had hidden in the henhouse because she wanted to watch a hen lay her egg. Fortunately for her, a year later—the start of the Second World War was imminent—the family sold their London home and, after a few stops in-between, moved to the country. Here Jane began to study the behaviour of local birds and other animals, and she was soon mature enough to record her observations in notes and sketches. Decades later, she recounts:

> I [wanted to] grow up, move to Africa, live with wild animals and write books about them. Everybody laughed at me. 'How will you do that?' they said. 'You don't have money. Africa is far away. It's a dangerous place. And, anyway, you're just a girl. Girls don't do that sort of thing.'[2]

Only *one* person encouraged Jane to make her dreams come true: her mother, the author of biographies and novellas Margaret Goodall. She always reacted calmly to earthworms in her daughter's bed and nocturnal escapades in the chicken coop:

> Jane, if you really want something, and if you work hard, take the chances and never give up, you will somehow find a way.[3]

When Jane was twelve years old, her parents divorced. The family could no longer afford to pay for her to study after finishing school. At the age of 18, she thus began an apprenticeship as a secretary and worked for various employers after completing her exams. For a while she was responsible for typing and filing at Oxford University. Here she felt very close to her dream of studying and going to Africa as a scientist, and yet at the same time infinitely far away. Jane Goodall was 23 years old when a stroke of luck gave her life a decisive turn that saved her from the fate of so many others who, because of their gender, lack of financial resources, and other circumstances, have no opportunity to fulfil their potential.

*   *   *

---

[2] Neale McDevitt, "*It's Been a Long Journey*", McGill Reporter, 1 October 2019, read at: https://reporter.mcgill.ca/its-been-a-long-journey-jane-goodall-tells-beatty-lecture-audience/.

[3] *Biography*, Jane Goodall Institute UK: https://www.janegoodall.org.uk/jane-goodall/biography.

In 1957, a former school friend invited Jane Goodall to visit her family's farm in the highlands of Kenya, which was still a British colony at the time. To finance the longed-for trip to Africa, Goodall worked as a waitress in the evenings in addition to her full-time job as a secretary. After five months, she had saved enough money for the boat trip to Mombasa, the largest port city in Kenya. After the end of her holidays on her friend's farm, she took up a job as a secretary in Nairobi, later the capital of Kenya, which she had already organised from England.

Africa! Jane Goodall had realised part of her dream. But she wanted more than to work as a secretary all her life. On the advice of a chance acquaintance, she made contact with archaeologist and palaeontologist Louis Leakey, who worked at the Museum of Natural History in Nairobi. Leakey was an expert on the animal life of Africa; his discoveries of fossils of human ancestors made him world famous. From the fossilised bones he could tell a lot about the physique and way of life of early humans. But how did they behave? Footprints in stone hardly allowed any conclusions about how they interacted in a group. Leakey wanted to gain insights into the behaviour of our extinct ancestors by studying modern apes—the chimpanzees of the sub-Sahara, the gorillas of Central Africa, and the orangutans of Southeast Asia. He was particularly interested in chimpanzees because they were the closest relatives of the genus *Homo* and their habitat also corresponded roughly to that of early humans.

His previous research trips with many participants had been failures. The animals' behaviour had obviously been influenced and the researchers had also spent too little time in the field to gain the animals' trust and thereby obtain insights into their natural behaviour. To eliminate these methodological flaws, Leakey had been searching since 1946 for a scientifically trained primatologist who was willing to endure a long-term research stay in the wild on his own. After ten years of fruitless searching, Leakey was ready to lower his standards for the formal training of such a person. In 1956, for lack of an alternative, he sent his secretary Rosalie Osborn to Mount Muhabura in Uganda to observe gorillas. But Osborn returned after four months and left Africa for England immediately afterwards. Leakey considered taking on the task himself—but just at that moment, the enthusiastic Jane Goodall appeared on the scene.

* * *

Leakey showed Goodall around the museum and tested her knowledge. He was enthusiastic about the young woman and invites her to accompany him and his wife and fellow researcher, the British palaeoanthropologist

Mary Leakey, to the Olduvai Gorge in present-day Tanzania as an assistant. There, where the Leakeys made their most significant discoveries, Leakey's first impression was confirmed: Goodall was a fearless, hard-working young woman with a good eye for detail. The latter might well have been due to her inability to remember faces and an unconscious attempt to compensate for this disadvantage by paying special attention. Goodall only learnt that her weakness had a name, prosopagnosia, decades later from the famous neurologist Oliver Sacks, who also suffered from it.

Goodall was thus exactly the right candidate for chimpanzee research. Then something happened that could easily have destroyed their relationship: in 1959, Louis Leakey fell in love with Goodall, who was more than 30 years younger, but she firmly rejected the affair. The two managed to preserve their friendship and steer their collaboration back into safer waters.

Leakey's plan to entrust Jane Goodall with chimpanzee research was met with incomprehension and rejection by the experts. Goodall had never studied. And anyway, how was a woman supposed to manage on her own in the wilderness? But Leakey trusted in Goodall's motivation and intelligence. He sent her back to England at the end of 1958, where she worked at London Zoo for over a year, observing the behaviour of captive chimpanzees in her spare time. This was her first step on the way to becoming an ape researcher. Later, she became the first of three women, subsequently called the "Trimates" or "Leakey's Angels", who would study the behaviour of apes in their natural environment:

- Jane Goodall (*1934) researched the behaviour of chimpanzees in the wild for three decades from July 1960. She was the first person to demonstrate behaviour such as personal affection and armed conflict in great apes. Today, researchers from various disciplines and nationalities still live and work with the direct descendants of the chimpanzees whose behaviour Jane Goodall first studied.
- The American Dian Fossey (1932–1985) spent twenty years in the Central African country of Rwanda from 1966 onwards researching the behaviour of gorillas. She campaigned vehemently for the protection of this ape species and fought hard against poaching in their habitats. In December 1985, she was murdered in a remote camp.
- The Canadian Birutè Galdikas (*1946) studied orangutans in Borneo for more than forty years from 1971 onwards and continues to campaign for the conservation of their habitat, which is severely endangered by slash-and-burn, logging, and resource extraction. In her memoirs *Reflections of*

Eden, she describes her work to free captive orangutans and release them back into the rainforest.

The example set by these three women has led to a situation where primate research, which was still a male-dominated field of research in the 1950s, is now carried out almost equally by men and women.

* * *

In 1960, the time had finally come. Jane Goodall, now 26, set off for the jungle. Her destination was the Gombe Reserve on the eastern shore of Lake Tanganyika, almost 1000 kms southwest of Nairobi. There, in the rainforest, about 150 chimpanzees were living in an area of about 25 square kilometres, characterised by steep slopes and basically more or less impenetrable. Jane Goodall's project was not without its dangers, and those responsible in the British administration insisted that she should not camp alone in the jungle. A companion was quickly found: her mother, who had always encouraged and supported her, and quickly got involved in the adventure during the first three months of the research.

In Gombe, Jane Goodall's determination and patience certainly paid off, because it would be many weeks before the chimpanzees stopped immediately disappearing whenever she approached. Even after almost a year, most of the apes wouldn't let her get closer than ten metres. And in the end, it took two years of daily observation and patient approaches before Goodall was finally accepted by the chimpanzee group and could move freely among them.

Over time, she succeeded in gaining many new insights into chimpanzee behaviour:

- For a long time, it was thought that only humans were smart enough to make and use tools; this ability was even seen as a clear dividing line between humans and animals. But Goodall observed a chimpanzee fishing for ants with a blade of grass in a rotting tree trunk. Later, it even bent a twig from a bush, removed its leaves, and used the little stick as a fishing rod, so it had not only used a tool, but also made one. Louis Leakey was thrilled and sent her a telegram:

  Now we must redefine tool, redefine Man, or accept chimpanzees as humans![4]

---

[4] Jane Goodall, *Learning from the Chimpanzees: A Message Humans Can Understand*, Science, 282 (1998), pp. 2184–2185.

Goodall's discovery opened the floodgates. In the years that followed, many other animal species were identified that used tools. In some, the behaviour had been known for a long time, for example, it was known that sea otters break open shellfish with stones to be able to eat their insides. It just hadn't occurred to people to see this behaviour for what it really was: the use of tools.

- Goodall saw a chimpanzee gnawing on a bone. The fact that apes eat meat was previously unknown to the experts. They had always been considered pure vegetarians.
- Goodall even discovered that chimpanzees go out specifically to hunt. She observed how a hunting group cut off a stub-tailed monkey from possible escape routes and killed it. Its meat was shared within the group. Years later, other researchers were able to show that the chimpanzees of Gombe kill up to a third of the mute-tailed monkey population every year. Later, they even observed chimpanzees sharpening sticks with their teeth and using them as spears to avoid being bitten by their prey.

The scientific world was in turmoil. Goodall's observations showed that chimpanzees were more similar to humans than had been previously assumed. Some behavioural scientists complained that Goodall was not a serious professional. They argued that the fact that she gave names to the chimpanzees in the group she was observing—Flo, Frodo, David Greybeard, Goliath, and many more—and even named them in this way in her publications instead of numbering the apes as was usually done, was evidence of a lack of objectivity and an unscientific way of working. In 2018, Goodall said in an interview:

(...) many did not believe me at the time because I was young and had never attended university.[5]

But Goodall's approach was anything but sentimental. She had recognised that each chimpanzee had a unique personality, with its own unique intelligence and individual traits. Some were more curious and willing to take risks, others more fearful and reserved. The degree of aggressiveness, caring, and many other characteristics also differed from one ape to another. The distinct social behaviour of chimpanzees within their own group was also more

---

[5] Interview with Liesa Bauer, published in: Spektrum Kompakt, issue 50/2019: *Be-/Verkannt - Frauen in der Wissenschaft*, available at www.spektrum.de/news/jane-goodall-ein-leben-fuer-die-schimpansen/ 1545469.

similar to humans than anyone had imagined. Goodall's well-documented observations were unequivocal:

- Chimpanzees share with each other, help each other, and show compassion. The fact that emotions such as joy, hatred, and sadness can be attributed to them was a completely new idea at the time.
- Chimpanzees exhibit behaviours such as hugging, kissing, patting on the back, and even tickling.
- Mothers and their children as well as siblings develop deep, often lifelong bonds with each other. When the mother dies, the younger offspring are adopted by the older siblings.
- If orphans have no close relatives to protect them, other families take them in.
- Goodall observed that the offspring of particularly attentive and playful mothers behave less depressively and aggressively than the offspring of other mothers.

The diversity of human behaviour—from selfless caring to deadly aggression—thus seems to be deeply rooted in our family tree. But Goodall's surprising research findings also had an unforeseen effect: the similarity of human and chimpanzee behaviour challenges the image of an unbridgeable gulf between us and them. The belief that humans are "the crown of creation" had already begun to falter with Darwin. Goodall brought humans and animals even closer together. In the early 2000s, the high degree of similarity was also shown at the genetic level: 98.5% of the genes of humans and chimpanzees are the same.

Goodall herself also had to say goodbye to a few preconceived ideas. She realised that chimpanzees not only hunt and kill strategically, but that massive violence also occurs *within* chimpanzee groups. For example, dominant females sometimes kill the young of other females in the group to maintain their status. Goodall even witnessed cannibalism. Once again, chimpanzees turned out to be closer to humans than anyone had previously thought: when competing for resources or under the influence of emotions such as jealousy, fear, and vindictiveness, they can behave as aggressively as humans. Goodall says about these relatively late discoveries:

For the first ten years of the study, I had believed that the Gombe chimpanzees were, for the most part, nicer than humans. Then we suddenly found out that chimpanzees could be brutal – that they, like us, had a dark side to their nature.[6]

In her book "Through a Window: My Thirty Years with the Chimpanzees of Gombe", published in 1986, she describes the details of the Gombe chimpanzee war from 1974 to 1978, a violent conflict that broke out between two chimpanzee groups once united in a community called Kasakela. Part of this group had split off and was subsequently renamed the Kahama community. The new group consisted of six adult males, three adult females and their young. Eight adult males, twelve adult females and their cubs remained in the original Kasakela group. In the course of the four-year conflict, all male members of the new Kahama community were killed and the community disbanded. The victorious Kasakelas expanded their territory, but were later pushed back by another chimpanzee community.

Comparisons with warlike confrontations among humans are obvious. The fact that male chimpanzees are far more aggressive than female group members is also reminiscent of *Homo sapiens*.

* * *

After Goodall's first two years in Gombe and her first surprising and significant observations, Leakey encouraged her to do a PhD in behavioural science at Cambridge. Although she had never attended university, she received the necessary exemption due to her significant research results. In the following years she worked—mainly in Gombe—on her dissertation *Behaviour of the Free-Ranging Chimpanzee*, receiving her doctorate in 1965. Now the wind was taken out of the sails of the critics who had denied her academic abilities.

Before that, the criticism already mentioned that Goodall was anthropomorphising the chimpanzees' behaviour and that her interpretations would not stand up to rigorous scientific scrutiny had also been dispelled. In 1962, the renowned *National Geographic Society* had sent the Dutch filmmaker Baron Hugo van Lawick to Gombe to document Jane's sensational work. The article *My Life Among Wild Chimpanzees,* which appeared in the *National Geographic* magazine in August 1963, became a resounding success. From this point, Goodall also became known to a wider public.

The documentary made for the *National Geographic Society* had another consequence: Goodall and van Lawick fell in love and married in London in March 1964. The former secretary Jane Goodall became Baroness Jane van Lawick-Goodall. The film *Miss Goodall and the Wild Chimpanzees,* directed

---

[6] Jane Goodall, *Through a Window: My Thirty Years with the Chimpanzees of Gombe,* 1986, now available from: Mariner Books (2010).

by Hugo van Lawick, was released in 1965, but not without some quarrels. This was because Goodall was appalled at how many errors were contained in the version intended by the producers. In order to push through a scientifically acceptable version, she hired a lawyer; and Orson Welles, who had been brought in as the narrator, had to voice the film again with the improved text. The film was a huge success, with 25 million viewers in North America alone. Jane Goodall was now famous, but above all her financial resources were significantly better.

Soon, critics were speaking out again. They were now complaining that Goodall was more of a film star than a serious researcher. But Goodall was relaxed about this:

> The media produced sensational articles, emphasising my blonde hair and shapely legs," she says now. "It didn't really bother me because that wasn't anything unusual." Then, revealing a resourceful pragmatism, she adds, "Anyway if my legs helped me to get publicity for the chimps – well, that was useful.[7]

In 1967, a son Hugo Eric Louis was born to the couple. Jane Goodall took him with her on observation tours, because this was what she had learned from the chimpanzees: the closer the contact between mothers and their offspring in the first years, the more balanced and stress-free the children grow up.

In 1968, the Gombe Reserve becomes the Gombe Stream National Park. The chimpanzees in this area were now largely protected from poaching and habitat loss. But the carefree days of the van Lawick–Goodall family and their joint research and documentation would soon come to an end. The *National Geographic Society* stopped funding van Lawick's work in Gombe and he had to spend a lot of time travelling again. The couple drifted apart and divorced in 1974. However, they remained good friends and continued to develop joint projects.

A year after the divorce, Goodall married the Tanzanian Derek Bryceson, who effectively supported her work as head of Tanzania's national parks. With his help, Goodall succeeded in founding the Jane Goodall Institute in Gombe in 1977. Its aim was to serve the further research and protection of great apes and their habitats and is today a globally active and influential non-profit organisation. Its nineteen branches are widely recognised for their conservation and development programmes.

---

[7] Available at: www.radiotimes.com/tv/documentaries/jane-goodall-if-my-legs-helped-me-get-publicity-for-the-chimps-well-that-was-useful/.

After only five years of marriage, Bryceson died of cancer in October 1980, aged only 58. Jane Goodall was 46 years old.

* * *

Goodall wanted to use her popularity as effectively as possible to preserve the chimpanzees' habitat, because population pressure and economic interests meant that large-scale deforestation of the rainforest was moving ever closer to the park. Where once pristine forests had stood, bare hills now stretched far and wide. Through her marriage to Bryceson, she had access to political decision-makers in Kenya and Tanzania and worked hard to make the population's encroachment on nature and the burgeoning tourism industry ecologically compatible.

> We had a session on conservation, and it was shocking to see right across Africa, wherever chimps were being studied, forests were disappearing. Chimp numbers were dropping, [it] was the beginning of the bush meat trade; chimpanzees caught in snares, mothers shot to steal babies for pets, for medical research, circuses and so forth.[8]

She had less and less time for her field research in Gombe. At the age of 52 and after more than 25 years on the steep slopes above Lake Tanganyika, she stopped her direct work with chimpanzees. This was triggered by her participation in an international conference in Chicago in 1986, where chimpanzee researchers from many disciplines had gathered to exchange ideas. Goodall watched film footage and listened to lectures documenting the excessive professional hunting of chimpanzees and the conditions in medical research facilities. She was appalled by the level of brutalisation with which humans were treating their closest relatives.

Goodall lectured around the world, raised funds, and set up new institutions dedicated to chimpanzee conservation. Here are a few examples:

- In 1984, *ChimpanZoo* was founded. This was an international research programme set up by the Jane Goodall Institute, dedicated to studying chimpanzees in captivity and improving their lives.
- In 1992, she founded the Tchimpounga Chimpanzee Rehabilitation Centre in the Republic of Congo, which consisted of three islands. It provided space for more than a hundred orphaned chimpanzees, which

---

[8] Oral statement by Jane Goodall at the *Understanding Chimpanzees conference* hosted by the *Chicago Academy of Sciences*, held in Chicago in 1986. Quoted here from: https://edition.cnn.com/2017/01/17/africa/jane-goodall-conservation/index.html.

were taken in and prepared for their release into the wild. It has been affiliated with the Jane Goodall Institute since 1999.

- In 1994, Goodall started the pilot project *Lake Tanganyika Catchment Reforestation and Education* (TACARE, also known as "Take Care"). Its aim was to protect the chimpanzees' habitat, organising the reforestation of the hills around Gombe and the education of the inhabitants of the surrounding communities in sustainable agriculture.

Goodall's focus was widening. It was not only the chimpanzees that needed support. She called for consequences in accordance with her research findings. In her book *Through a Window*,[9] she describes why knowledge about the mental and social complexity of animals must lead to responsible ways of dealing with them. Keeping animals as pets, for entertainment, in experimental laboratories, and especially for meat production had to be fundamentally reconsidered.

Goodall launched numerous projects including:

- In 1991, an informal meeting with 16 teenagers on the veranda of their house in Dar es Salaam, Tanzania, gave rise to the global programme *Roots and Shoots,* which encouraged and supported young people to get involved with people, animals, and the environment. At Goodall's instigation, thousands of projects in more than a hundred countries are now running under this umbrella.
- In 2000, together with the US biologist and ecology professor Marc Bekoff, she founded the organisation "Ethologists for the Ethical Treatment of Animals."

Even at the age of over eighty, Goodall travelled tirelessly around the world to represent her causes and raise money for her many and varied undertakings. Ever since she attended the Chicago conference at the age of 52, she has never allowed herself any rest:

(...) since that day, I haven't spent more than three weeks in any one place, except once when I tore the ligaments on both ankles, and I needed an extra week or so to get better.[10]

Only the Corona pandemic stopped Goodall's restless life.

---

[9] Jane Goodall, *Through a Window*, Mariner Books (1990).

[10] Interview with Alice Winkler for the *American Academy of Achievement's "What it takes"* podcast, 18 May 2018, read at https://learningenglish.voanews.com/a/what-it-takes-jane-goodall/4364308.html.

Numerous honorary doctorates and other honours bear witness to the high regard in which she is held, and indeed her commitments are held, all over the world:

- In 2002, she was appointed UN Messenger of Peace by UN Secretary-General Kofi Annan.
- In 2004, she was made a *Dame of the British Empire* at Buckingham Palace. This honour is equivalent to a knighthood; the title "*Dame*" is the feminine form of "*Sir*".
- In 2006, she became a member of the French Legion of Honour and received the UNESCO Gold Medal.

As a unique scientist, Jane Goodall went her very own way and gave behavioural research a completely new impetus. She drew the consequences from her findings and became politically active. Her unwavering attitude led to her scientific enthusiasm and tireless and detailed research naturally becoming an unconditional commitment to the welfare of animals and the preservation of their habitats. The fact that she became an icon of humanity is also due to another quality: her persistence in realising her dream. When she was once told in an interview that it had been very courageous to research the behaviour of chimpanzees alone in Africa, she replied:

It wasn't brave, it was my dream.[11]

---

[11] Read more at www.spektrum.de/news/jane-goodall-ein-leben-fuer-die-schimpansen/1545469.

16

Jocelyn Bell Burnell (*1943): Our Guide to a New Universe

Astronomy is probably the oldest of all sciences. People have been observing the course of the stars since time immemorial, and some of the earliest writings describe the processes going on in the night sky. Mathematics and astronomy have always fuelled each other, so Chinese scholars were able to predict solar eclipses more than 4000 years ago. The fact that the stars are not simply positioned on a hemisphere placed over the Earth's disk, but that each one is a sun in its own right, had been suspected by a few philosophers since antiquity, but it was only surprisingly late, at the beginning of the nine-teenth century, that this could be demonstrated scientifically. This was when spectroscopy was developed. This is a technique that separates incoming light into its colour components through a prism and also shows the absorption lines typical of each light source. It was shown that the light emitted by the Sun and the light emitted by stars have basically the same properties.

Spectroscopy opened new doors for astronomers. For many millennia, it had been assumed that the stars would forever elude our human thirst for knowledge. But now it was possible to assign a colour class and a surface temperature to each of them. In the following decades, the Harvard Observatory collected the spectral lines of more than 200 000 stars and brought them together in a catalogue. It was soon discovered that stars are not immutable, but are formed from agglomerating interstellar clouds and eventually collapse at the end of their lifetime due to constant radiation of energy. Depending on the mass they start out with when they first form, they end up as brown or white dwarfs, neutron stars, or black holes. Our own star, for example, will eventually evolve into a red giant before it ends up as a white dwarf.

The invention of the telescope by Galileo Galilei gave astronomy a tremendous boost, and it remained a science based solely on optics until well into the twentieth century. It was only after the Second World War that astronomers began to search the sky systematically, not only for visible light waves, but also for radiation of other wavelengths.

The Earth atmosphere is permeable to electromagnetic radiation with wavelengths of 360 to 830 nm, i.e., for the visible wavelength range. Evolution has accordingly produced living beings that are precisely tuned to this spectrum with their sensory organs (animals) and their systems for generating energy (plants). Fortunately for astronomers, it turned out that the atmosphere also has a window for electromagnetic radiation in the wavelength range of a few millimetres to 20 m. This is the radio wave range. Radio astronomy developed from the use of these wavelengths. Today, there is also gamma, X-ray, ultraviolet, and infrared astronomy, but this can only be done from outside the Earth's atmosphere, for example from satellites.

* * *

Because the wavelengths used in radio astronomy are absorbed less than light waves by interstellar matter, new insights into the structure of the universe became possible. In the middle of the twentieth century, radio telescopes were built all over the world, including in Cambridge, England. Here, Antony Hewish, a physics lecturer at Churchill College in Cambridge, built an 81.5-megahertz radio telescope with the help of his doctoral student Jocelyn Bell and other colleagues. The aim was to search for quasars. These celestial bodies had been discovered by radio astronomers in 1960. At first, they were thought to be special stars, but then it became clear that they were the active nuclei of galaxies, and in fact rather small objects, barely the size of our Solar System, yet sometimes emitting as much energy as hundreds of galaxies. Since their luminosity is up to 10^{12} times that of the Sun, they are often visible in optical telescopes despite their position at the confines of the universe.

Jocelyn Bell was technically gifted and had built a prototype of the Cambridge radio telescope. Based on this prototype, the actual radio telescope was built on about one and a half hectares of land. It took two years for Hewish, Bell, and the other workers to attach the almost 200 kms of wire that served as the radio antenna to scaffolding and solder the whole thing together in all weathers in a field near Cambridge. Jocelyn Bell was also involved in this manual work. Fifty years later, she remembers:

I helped build the radio telescope, along with about five others. And when it was built, the rest melted away. I was left as the first person to run the telescope.[1]

The radio telescope went into operation in 1967. Rather as we see from seismographs, the captured signals left a jagged line on a continuous printout. In the following months, Bell, a PhD student, had the task of operating the radio telescope and evaluating these printouts of the captured signals. Every four days, Bell collected the 120 m of printer paper accumulated during this time and evaluated the signal line at her workstation. She soon developed an eye for interference signals caused by people. When she couldn't classify a signal, she would put a question mark on the endless printout, or sometimes, just for fun, the initials "LGM", which stood for "little green men".

On 6 August 1967, she noticed for the first time a particular irregularity in the never-ending signal line: on the endless metres of the printout, it extended over just a few millimetres. Bell marked it with a question mark. The radio telescope scanned different parts of space thanks to the Earth's rotation, and on the following days, too, the tiny disturbance appeared whenever the radio telescope was pointed at exactly the same part of the sky again, 24 h later. Other scientists might have overlooked these tiny anomalies, but they gave Bell no peace. The regularity of their occurrence told her that it could not be a coincidence.

She decided to discuss her discovery with her doctoral supervisor Antony Hewish. He initially thought the disturbances were terrestrial radio interference, but Bell suspected the origin of the anomalies to be far outside the Solar System. She needed a lot of persuasion to get Hewish to continue working on the mystery. On 28 November 1967, the two succeeded in temporally resolving the anomalies clearly enough to reveal their structure: they were signal sequences consisting of individual bursts that occurred uniformly every 1.337301094 s. The astonishing precision of this pulse was on a par with that of an atomic clock. Bell and Hewish were baffled. Such an astronomical object had never been described before. Was this an attempt by an intelligent alien life form to make contact? Bell later said about this phase:

We did not really believe that we had picked up signals from another civilization, but obviously the idea had crossed our minds and we had no proof that it was an entirely natural radio emission. It is an interesting problem – if

---

[1] Jocelyn Bell's speech at the 2017 50th anniversary meeting; see also: www.nationalgeographic.com/science/article/news-jocelyn-bell-burnell-breakthrough-prize-pulsars-astronomy.

one thinks one may have detected life elsewhere in the universe how does one announce the results responsibly?[2]

Hewish did not rule out the idea that the signals might be from an extra-terrestrial intelligence, but did not want to make a fool of himself either. He used the abbreviation LGM introduced by Bell and called the still unknown source of the signal "LGM-1". By bringing the *little green men* into play in this joking way, he simultaneously distanced himself from the possibility. Bell was already convinced at this point that the signal had to have a natural cause. She set out to find more signals of this kind, and on 21 December, far from the first source, she discovered a second spot in the sky that was transmitting radio signals with extreme accuracy; shortly afterwards, even a third and a fourth. This put an end to the speculation that these could be contact attempts by living beings from an alien planet.

In February 1968, the renowned magazine Nature published the article "Observation of a rapidly pulsating radio source".[3] The first author was Antony Hewish, Jocelyn Bell was listed second. James Pilkington, Philip Scott, and Richard Collins, who were listed as further authors, had hardly been involved in the discovery.

The publication was met with great interest. The sources of the radio signals discovered by Bell were still a mystery. It was several months before astronomers identified the sources as neutron stars. Their existence had already been theoretically predicted in 1934 by the German astrophysicist Walter Baade and the Swiss astrophysicist Fritz Zwicky. They argued that a supernova could give rise to a star consisting mainly of neutrons. According to their theory, under certain circumstances, the pressure created during the collapse should be so great that electrons are forced into the atomic nucleus and combine with the protons present there to form neutrons. At the beginning of the 1960s, interest in neutron stars had been rekindled:

- In 1964, the Dutchman Lodewijk Woltjer argued that neutron stars must generate enormously strong magnetic fields due to the conservation of the intensity of a magnetic field (magnetic flux), because the speed of rotation would be much higher than before the collapse due to the small size of the star.

---

[2] Jocelyn Bell Burnell, *Little Green Men, White Dwarfs or Pulsars?* Annals of the New York Academy of Science, 302 (1977), pp. 685–689.

[3] Antony Hewish, Jocelyn Bell, James Pilkington, Philip Scott, Richard Collins, *Observation of a Rapidly Pulsating Radio Source*, Nature, 217 (1968), pp. 709–713.

- Just a few months before Bell Burnell discovered the first pulsating radio source in 1967, the Italian astrophysicist Franco Pacini had shown that a rotating neutron star must emit observable electromagnetic radiation.
- Shortly after the discovery of pulsed radiation, the US astronomer Thomas Gold was the first to make the connection with rotating neutron stars having a strong magnetic field. At the first conference dealing with Bell's discovery, however, his suggestion was still considered absolutely absurd.

But Gold's idea was spot on. Today we know that some stars whose mass at birth is significantly greater than the Sun's become neutron stars when they collapse. Since a lot of the original mass of the star gets concentrated in a sphere less than twenty kilometres in diameter, the density is enormously high. Conservation of angular momentum causes these "stellar corpses" to rotate extremely quickly around their own axis; pirouettes in figure skating are also based on this effect. Their rotation time is only seconds or fractions of a second. The direct result of this ultra-fast rotation are super-strong magnetic fields with a strength of 10^9 to 10^{11} T. By comparison, the Earth generates a magnetic field of about 60 microtesla, or 60 millionths of a tesla, and man-made magnetic fields of 20 to 30 T are considered very strong. Since the rotation axis of a neutron star never coincides exactly with its magnetic field axis (just as with the Earth), two so-called radiation cones are created at the magnetic poles, giving an effect rather like a lighthouse. Whenever one of them points towards Earth, very strong electromagnetic radiation is sent in our direction.

At the beginning of March 1968, the word "pulsar" appeared in the press for the first time. The term is an abbreviation of *pulsating quasar* and is used today for all pulsating radio sources.

> On 6 August last year, a completely new type of star was discovered, which astronomers named LGM (Little Green Men). Now it is believed to be a new type between a white dwarf and a neutron star. It will probably be given the name pulsar. Dr. A. Hewish told me yesterday, '...I am sure that today every radio telescope is looking for pulsars.'[4]

In fact, radio astronomers detected more than 2000 pulsars in the following fifty years. Of these, about two hundred transmit millisecond pulses, implying that they are very young neutron stars because the rotational energy decreases over time. About 150 pulsars are binary star systems. This high number of discoveries comprises only a tiny fraction of the neutron stars

---

[4] Daily Telegraph on 5 March 1968.

that must exist in theory. In our galaxy alone, there are probably more than 100 million of these surprising celestial bodies.

Thanks to their extraordinarily high density, pulsars can be used to test some of the most fundamental theories of physics, for example that of gravitational waves. In 1993, the American researchers Joseph Taylor Jr. and Russel Hulse received the Nobel Prize for the indirect detection of gravitational waves emitted by pulsars. This was done with the help of long-term measurements of the orbital periods of a neutron star double system and comparison of the data with the pulsed radiation. It turned out that the observed convergence of the two neutron stars corresponded perfectly to the loss of total energy from the system if it were emitting gravitational waves.

The pulsars discovered by Jocelyn Bell thus proved to be a real treasure trove for astronomers. For a young PhD student to make such a significant discovery is extremely rare in the history of science. Was it a chance find that could have been made by anyone who was in the right place at the right time? Or was there more to this achievement?

* * *

Jocelyn Bell was born on 15 July 1943 in the Northern Irish town of Armagh near Belfast. Her mother Margret had not been able to go to school, as her family could only give her brother an education for financial reasons. This experience led Margaret Bell to place particular emphasis on the education of her three daughters. Their father Philip Bell was an architect and wealthy. Jocelyn Bell Burnett recalls:

> We had a nanny, a nurse always. For a long time there was a cook and a maid in the kitchen. There was a gardener and maybe a second gardener and there was a handyman-come-chauffeur if you want to be really grand. It was a great life.[5]

Both parents were Quakers. This religious community, founded in England in the seventeenth century, rejected the clergy and their customarily elaborate ceremonies from the very beginning. With their pacifism, their rejection of hierarchies, and their conviction that all people without exception are equal, the quakers made no friends in English society. As a persecuted community, they were among the first settlers in America, where they distinguished themselves as vehement opponents of slavery and supporters of the women's movement, among other things.

---

[5] Conor Moloney, *Jocelyn Bell: The True Star,* Belfast Telegraph on 3 July 2008, available at: www.bel fasttelegraph.co.uk/life/jocelyn-bell-the-true-star-28528404.html.

Jocelyn Bell grew up in a family that lived these values. While it was still the custom in her environment for boys to find the whole world open to them and for girls to learn sewing and cooking, her parents encouraged their daughter's interest in science. The family often visited the planetarium in Armagh, which was designed by her father, so Jocelyn came into contact with astronomy as a child. Her father's astronomy books probably also contributed to her desire to learn more about this science.

In the English school system, primary school continues to the age of eleven. When Jocelyn was eleven, she had to take an important exam that would decide which secondary schools were open to her. She did very badly, so her path to higher education was blocked. Jocelyn was crushed, but her parents knew a way out. They sent their daughter to a Quaker girls' boarding school in York, England, where she qualified for university in 1961. In 1965, Jocelyn Bell graduated with honours in physics. She was then offered a place at the prestigious University of Cambridge to study for a PhD. So Bell's educational journey had started in a rather mediocre way and ended at one of the most famous universities in the world. She later said:

> I was quite sure they'd made a mistake in admitting me, and I was sure they were going to throw me out at some point when they discovered their mistake. So, I decided my game plan was to work as hard as I could so that when they threw me out I wouldn't have a guilty conscience, I'd known I'd done my best.[6]

Massive self-doubt is surprisingly widespread among people who achieve the highest performance. This so-called impostor syndrome was only finally named at the end of the 1970s. But Jocelyn Bell Burnell was anything but an impostor. She was just 24 years old when she discovered the first pulsars in 1967.

* * *

In 1974, Hewish received the Nobel Prize in Physics for the discovery of pulsars. Martin Ryle, who also did research in Cambridge and had laid the foundation for Bell's discovery through his development of revolutionary radio telescope systems for the precise location and imaging of weak radio sources, was jointly awarded the prize. It was the first Nobel Prize in Physics to go to astronomers. The justification can be read on the official website Nobel Prize Commission:

---

[6] Lucie Edwardson, '*Go for it': Female Scientist Famous for Discovering Pulsars Encourages Girls to Pursue Science,* interview with CBC News on 19.9.2018; available at www.cbc.ca/news/calgary/pulsars-calgary-jocelyn-bell-1.4830811.

The Nobel Prize in Physics 1974 was awarded jointly to Sir Martin Ryle and Antony Hewish for their pioneering research in radio astrophysics: Ryle for his observations and inventions, in particular of the aperture synthesis technique, and Hewish for his decisive role in the discovery of pulsars.[7]

Jocelyn Bell went away empty-handed and was not even mentioned. Several prominent scientists protested against this decision. Bell later recounted that one of them had called the Nobel Prize a "no-Bell" award.[8] She herself found the decision understandable, as she had only been involved in the development as a doctoral student at the time of the discovery. At the same time, she saw her contribution as highly crucial to the discovery of pulsars. In her dissertation she wrote:

> If the data output from the antenna had been digitised and fed directly into a computer, these sources would probably not have been detected because the computer would not have been programmed to look for such unexpected objects.[9]

Hewish disagreed. In 2008, he made the following comments:

> I'm totally sick of it, this stupid thing that Jocelyn would have done all the work and I would have got all the credit. I kind of see an analogy there: Who discovered America? Was it Columbus or was it the man in the lookout? Her contribution was very useful, but not creative. And I don't think you get the Nobel Prize for that.[10]

One can argue whether Bell deserved the Nobel Prize. The more significant question is: Was Bell's discovery a "one-hit wonder"? Or does her accuracy and persistence make her a gifted scientist who hit the bull's eye early in her career and continued to do excellent science?

* * *

The fact is that Jocelyn Bell had to fight her way for a long time before she reached a status where it no longer mattered whether she was a woman or a

---

[7] Nobel Prize Committee 1974; www.nobelprize.org/prizes/physics/1974/summary/.

[8] Louise Walsh, *Journeys of Discovery*, interview at Cambridge University (2020); available at www.cam.ac.uk/stories/journeysofdiscovery-pulsars ((respect rights!)).

[9] Jocelyn Bell Burnell, *The Measurement of Radio Source Diameters Using a Diffraction Method* (PhD thesis from 1969); https://doi.org/10.17863/CAM.4926.

[10] Conor Moloney, *Jocelyn Bell: The True Star*, Belfast Telegraph on 3 July 2008, read at *Jocelyn Bell: The True Star* - BelfastTelegraph.co.uk; www.belfasttelegraph.co.uk/life/jocelyn-bell-the-true-star-285 28404.html.

man. At Glasgow University, she had been the only woman to study physics. In a 2011 interview with the Guardian, Bell Burnell recalls:

> It was distinctly tough. I ended up in the final two years of my course as the only female in the class, there were 49 men and me. There was a tradition among the students that when a female walked into a lecture theatre all the guys stamped and whistled and called and banged the desk. And I faced that for every class I walked into for my last two years.[11]

Even a few years later, when her discovery of pulsating radiation had become a media event, she was confronted with the expectations that society had of women in those days: Bell's doctoral supervisor Antony Hewish was asked about the astrophysical significance of the find, Bell about breast, waist, and hip size. And when photos were to be taken, she was asked to open a few buttons on her blouse. And she had to deal with such impositions for a long time to come.

> In those days it was believed that science was made and driven by big men, and that these men had a lot of foot soldiers under them who did whatever they wanted and had no thoughts of their own. There also came a time when I had a young child and was struggling to juggle finding suitable childcare and a career - all before it was acceptable for women to work. Now, men win awards and young women look after babies.[12]

Not only her professional career, but also her private life turned out to be anything but easy. In 1968, the year of her important publication and before she had even received her doctorate, Bell became engaged to the government official Martin Burnell and married him shortly afterwards. From then on, her name was Jocelyn Bell Burnell. More than half a century later, Bell Burnell recounted what happened in an online lecture[13] when, newly engaged, she shared the good news with her colleagues at the Cambridge Observatory. Instead of good wishes, there was criticism because Bell had no intention of giving up her job. Even in 1968, it was considered embarrassing for a married woman to pursue a job, because this gave the impression that her partner was unable to provide for the family.

Jocelyn's husband Martin Burnell had to move several times for professional reasons, and each time she followed him to the new location. Yet, she nevertheless managed to advance her career:

---

[11] Ibid.

[12] Interview on BBC (25 October 2011); www.bbc.co.uk/sounds/play/b016812j, after 1 min 35 s.

[13] Online lecture on 3 June 2021, Fourth Edwards Lecture, *University of London.*

- After completing her doctorate in 1969, Bell Burnell taught as a junior professor at the University of Southampton from 1970 to 1973. Here she worked on gamma-ray astronomy, which—unlike radio astronomy—deals with particularly short-wavelength, high energy radiation. Among other things, she worked on the development of a gamma-ray telescope for energies of 1 to 10 million electronvolts (MeV).
- From 1974 to 1982, she researched and taught in the field of X-ray astronomy at London University, and from 1982 to 1991, infrared astronomy at the *Royal Observatory* in Edinburgh. In this way, she gained experience working with many different wavelengths.
- In 1973, her son Gavin was born and Bell Burnell worked only part-time for many years. In the same year, she received the Albert A. Michelson Medal from the *Franklin Institute of Philadelphia* in the USA (now called the Benjamin Franklin Medal).
- From 1973 to 1987, she was a tutor, advisor, examiner, and lecturer at the renowned Open University in Milton Keynes. This distance learning university is the largest university in the UK in terms of student numbers.
- In 1978, she received the *Robert Oppenheimer Memorial Prize* for theoretical physics, which came with a gold medal and a thousand US dollars.
- In 1986, she became head of the department responsible for the James Clerk Maxwell Telescope on Mauna Kea in Hawaii.
- In 1989, she received the Herschel Medal of the *Royal Astronomical Society*.

All the positions and honours, only some of which are listed here, bear witness to the fact that Bell Burnell was already perceived as an excellent astronomer in the first half of her career. Her husband, however, did not cope well with her success.

> When I got awards, my husband wasn't really as pleased as I thought he would be, and I learned not to make a big fuss about it when I came home with the news that I was getting an award.

The marriage soon began to come apart. In the end, Martin Burnell left his wife in 1989. After the divorce in 1993, Jocelyn Bell Burnell kept her name. She then had to fight her way through an emotionally and financially strained time, but courageously persevered with her career. Here are just a few of her other key achievements:

- In 1991, she became a professor of physics at the Open University, where she had already been a lecturer since 1973.

- In 1999, she became *Commander of the Order of the British Empire.*
- From 2001 to 2004, she was Dean of Natural Sciences at the University of Bath. During this time, she was also a visiting professor at Oxford and Princeton.
- She was President of the Royal Astronomical Society from 2002 to 2004.
- In 2003, she became a member of the Royal Society.
- In 2007, Queen Elizabeth II ennobled her as a *Dame of the Order of the British Empire.*
- She was President of the Royal Society of Edinburgh from 2014 to 2018.
- In 2015, she received the *Royal Medal,* which is awarded for particularly important contributions to science.
- In 2021, she received three important awards: the Gold Medal of the Royal Astronomical Society, the Copley Medal of the Royal Society, and the Karl Schwarzschild Medal of the German Astronomical Society.

In March 2022, Northern Ireland's Ulster Bank unveiled its new £50 banknote featuring Jocelyn Bell. The bank describes her as a crucial contributor to the advancement of science.

Among Bell Burnell's most significant honours is the *Special Breakthrough Prize*, awarded at irregular intervals, which she received in 2018 for her life-long inspirational leadership in science, and explicitly for her 1967 discovery of pulsars—so this can be considered as a late tribute to her very first scientific achievement. This most prestigious science prize has a monetary value of three million US dollars, more than double the amount awarded to Nobel laureates. Even greater is the honour associated with it: Bell Burnell now stands beside physicists like Stephen Hawking and the discoverers of the Higgs boson.

In keeping with the tradition of her Quaker faith, Jocelyn Bell Burnell decided to use the prize money to establish the *Bell Burnell Graduate Scholarship Fund*. This foundation supports female students, members of minorities, and refugees to obtain a doctorate in one of the many research fields of physics. Outside of this endowment, Bell Burnell is also a strong advocate for diversity in science and a voice that is increasingly heard.

17

Lisa Randall (*1962): High-Flyer in Theoretical Physics Today

Ever since it became clear that Newtonian physics can only be applied on the scale familiar to humans, scientists have been working on developing theories for the very small scales of the subatomic world and the very large scales of space that can be used to record and calculate the respective processes.

- The theory of general relativity developed by Albert Einstein at the beginning of the twentieth century describes how energy and matter distort the four-dimensional space–time and cause gravity. This theory can be successfully applied to the very large size scales of planetary orbits and galaxies, but not to the subatomic world.
- Quantum mechanics, developed around 1925, and quantum electrodynamics, developed in the 1940s, describe the phenomena of the subatomic world very well. In the following decades, these theories were improved and unified into the so-called Standard Model of particle physics. It works with three of the four basic forces of physics: the strong and the weak nuclear force as well as the electromagnetic force. The fourth basic force, gravity, does not fit in here.

The Standard Model of particle physics and general relativity work very well in their respective domains. But so far, physicists have not succeeded in unifying the three basic forces of the Standard Model into a single force. To demonstrate this so-called Grand Unified Theory (GUT), they would have to reach the exceedingly high energy of 10^{15} giga electronvolts (GeV)

© The Author(s), under exclusive license to Springer Nature Switzerland AG 2023
L. Jaeger, *Women of Genius in Science*,
https://doi.org/10.1007/978-3-031-23926-7_17

in experiments. The most powerful particle accelerators today achieve just 7 tera electronvolts (TeV), i.e., $7 * 10^3$ GeV. So, there are still 11 orders of magnitude to go before they achieve the necessary energies.

There is another reason why the Standard Model of particle physics is anything but satisfactory: it is complicated and inelegant, and many scientists view it as simply ugly. Einstein's general theory of relativity is much more to their liking. His famous formula $E = mc^2$, which derives from his special theory of relativity, is also viewed as being of the greatest beauty.

Even more ambitious is the desire of physicists to find an overarching, elegant theory that unites GUT and general relativity. This *Theory of Everything* (TOE) would describe subatomic particles just as well as stars and galaxies and, what's more, show that all four basic forces acting between them spring from a common denominator.

TOE is a kind of holy grail for physicists. But so far, attempts to unify the Standard Model and general relativity into this common theory all seem to have failed. If, for example, gravity is formulated at the quantum level to calculate processes occurring in the big bang or in black holes, the equations go haywire. Infinite values appear that make any attempt at calculation impossible. It is therefore assumed that TOE can be developed neither from general relativity nor from today's Standard Model, but requires a completely new approach.

* * *

One of the specialists in search of the big picture is the theoretical physicist Lisa Randall. As a leading expert in the fields of both particle physics and cosmology, i.e., the study of the universe as a whole, she brings with her the best prerequisites for packing both the phenomena of the very small and the very large scales into a common model. She has great respect for this goal, because she is well aware of the limited possibilities available to humans, at home as we are only at intermediate length and time scales, to be able understand the universe. In an interview with the historian of science Harald Zaun, she said:

> You know, cosmologists are not really looking for the world formula, because in that respect most answers are beyond their reach. Cosmologists are looking rather for what they can see.[1]

---

[1] Translated by LJ; Read more at www.astronomie.de/bibliothek/interviews/professor-lisa-randall-2006/.

Despite or precisely because of this modesty, she has been very successful in her search for universal connections. In addition, there is another important quality without which she would not be able to grasp the complexity of mathematics in cosmology and particle physics: she is a genuine achiever. Her ingenuity and flexibility run like a thread through everything she does.

Lisa Joy Randall was born on 18 June 1962 in Queens, a borough of New York, the second of three daughters. Her father was a sales representative, her mother an elementary school teacher. Lisa attended a science-oriented high school in Manhattan, from which she graduated in 1980. In the same year, she won first place in the *Westinghouse Science Talent Search*. This competition for high school graduates is also known as the "Super Bowl of Science." After that, things went from strength to strength. With a scholarship, Randall went to Harvard in 1980, where she studied physics and graduated with a bachelor's degree in 1983. Even at this early stage, she was concerned with fundamental questions of theoretical physics. In her doctoral thesis *Enhancing the Standard Model*, she worked on improving the Standard Model of particle physics. In 1987, she received her doctorate at the age of 25.

She then moved from the East Coast to the West Coast of the USA. In Berkeley, she first worked at the University of California, then from 1989 at the *Lawrence Berkeley National Laboratory*. In 1990, she returned to Harvard, where she quickly climbed the career ladder:

- In 1991, she became an assistant professor.
- In 1995, she reached the next professorial level at the Massachusetts Institute of Technology (MIT).
- In 1998, the 36-year-old Randall became the first woman to be appointed to the chair of theoretical physics at Princeton. At the same time, she was a full professor of physics at MIT, only the second woman to hold a chair in the physics department there.
- In July 2001, when Randall was not yet 40 years old, she moved once again, this time back to Harvard, where she now holds a chair in theoretical physics.

Harvard, Princeton, and MIT in Cambridge, i.e., three of the most important and prestigious universities, were all vying for Lisa Randall. So, at the end of the 20th and beginning of the twenty-first century, it had finally become possible—though still far from self-evident—for women to make a career in science. While Randall's predecessors were ridiculed and cold-shouldered by their male colleagues despite their excellent achievements, Lisa Randall managed to make it to the very top of research in almost record time.

Let us consider for a moment her contribution to current research. Her name is often mentioned in connection with the Randall–Sundrum model. Since this is a development of string theory, the term will be explained below. The whole story begins with people looking beyond the three-dimensional space we are familiar with:

- Some mathematicians had already toyed with the idea that time was like a fourth dimension in the eighteenth century. Einstein's special and general theories of relativity introduced four-dimensional space–time into physics extended this tradition of thought.
- For the mathematician Bernhard Riemann, there were only spatial dimensions. In the middle of the nineteenth century, he worked out a geometry in which spaces could assume any number of dimensions. What many mathematicians initially perceived as a superfluous gimmick later turned out to provide the indispensable basis for much higher mathematics and theoretical physics.
- The English mathematician and philosopher Charles Hinton understood the fourth dimension as an extension of space, and in 1880 he described a four-dimensional cube that he called a tesseract, a term that has haunted science fiction literature ever since.

Incidentally, at the end of the 19th and beginning of the twentieth century, other parts of society were also fascinated by the idea that there could be more than just three dimensions. Pablo Picasso's cubist portraits are an example of the hype that prevailed at the time. These are the results of his attempt to paint from different perspectives at the same time, as if he were a being from the fourth spatial dimension and could perceive all angles simultaneously.

Slowly, science came to realise that there could be no world-explaining theory that uses only three spatial dimensions. Higher-dimensional models alone could offer a solution. It was as if living beings who inhabit a two-dimensional world were trying to explain why a disk they observe grows larger and then smaller again in regular cycles. This phenomenon is an unsolvable riddle for them until someone comes up with the idea of adding another dimension to the two-dimensional world. Only then do they realise that it is a three-dimensional sphere bouncing up and down in a dimension that is invisible to them, but which intersects their two-dimensional plane.

Mathematicians continued to hone their ability to calculate with many dimensions, and at the end of the 1960s, they found a method with which theoretical physicists could finally unite the four basic forces of physics in a theory valid for all orders of magnitude. It was completely crazy: if they assumed a 26-dimensional space–time, the standard theory of particle physics and general relativity could be combined, because in this space, the infinitely high values resulting from their calculations, which had otherwise always led to a dead end, could finally be circumvented.

This theory of 26-dimensional space was given the name string theory, because the fundamental trick was to replace zero-dimensional point particles with tiny rolled-up one-dimensional threads called strings. The Swedish theoretical physicist Oskar Klein had already devised the decisive trick in 1926. For many decades, his description had been considered a useless mathematical exercise. But a good forty years later, physicists began to imagine the building blocks of the world as vibrating strings many billions of billions of times smaller than an atomic nucleus.

String theory with 26 dimensions had an appealing logic and could solve a whole series of problems for theoretical physicists. But elsewhere, new inconsistencies emerged. For example, it could not explain the existence of a certain type of quantum particle. In 1971, string theory was extended, by the introduction of another mathematical hypothesis, to superstring theory. In this theory, space–time is only ten-dimensional. Further string theories followed, which work with eleven dimensions. But even these extensions are still far from being a convincing TOE.

* * *

String theories thus work with several dimensions in order to be able to explain the phenomena that we perceive in the four-dimensional space–time we are familiar with. This is where Lisa Randall comes in. As an expert in string theories, she and Raman Sundrum, a US theoretical particle physicist born in Madras, India, developed the Randall–Sundrum model in 1999. This has some amazing properties[2]:

- The four-dimensional space–time in which we live is embedded in a five-dimensional universe.
- Other four-dimensional spacetimes, i.e., parallel universes, can also be embedded in this five-dimensional universe.
- Because the five-dimensional universe is tightly curled up, we are—probably—insurmountably separated from these parallel universes.

---

[2] Lisa Randall and Raman Sundrum, Large Mass Hierarchy from a Small Extra Dimension, *Physical Review Letters*, 8 (1999).

The Randall–Sundrum model offers answers to many problems that theoretical physicists have encountered. Even one of the greatest puzzles of physics can be explained with it: the so-called hierarchy problem. It concerns the Higgs particle. This is what gives the massive elementary particles their mass in the first place, and it is therefore also responsible for gravity. According to standard theory, the Higgs particle should actually have many orders of magnitude more mass. Although the gravitational force can keep planets and galaxies in their orbit, at the subatomic level it is surprisingly weak compared to the other three basic forces:

- Despite its name, the weak nuclear force in a hydrogen atom consisting of a proton and an electron is 10^{26} times stronger than gravity. For comparison, there are about 10^{24} stars in the entire universe.
- The electromagnetic force that binds the proton and the electron together in the hydrogen atom is 10^{39} times stronger than their gravitational attraction.
- The strong nuclear force is stronger than gravity by a factor of 10^{41}.

In the Randall–Sundrum model, these three forces are limited to "our" three dimensions, while gravity also penetrates the other two dimensions and thus loses a large part of its force.

Such models stem from a great imagination. But pure imagination does not make a theoretical physicist. With the most complex mathematics, some of which has yet to be developed, the ideas must be presented in a coherent manner. The fact that in the Randall–Sundrum model only gravity, but not the three other forces, can leak into the two "supernumerary" dimensions is not arbitrary; it is based on incorruptible mathematics. Lisa Randall put it this way:

> Our idea was this. If space still has a tiny, coiled-up dimension – we call it a brane, derived from membrane, then gravity could act much more strongly in this brane than in the remaining dimensions. In this case, the low Higgs mass arises quite naturally.[3]

Another advantage of the five-dimensional Randall–Sundrum model is that it could be demonstrated experimentally. To demonstrate the phenomena predicted by the other string theories, enormously high energies would be needed. Given today's technology, a particle accelerator would have to be as large as our galaxy to confirm those string theories. For the Randall–Sundrum

---

[3] Translated by LJ. Read more at https://science.orf.at/v2/stories/2874349/.

model, on the other hand, the Large Hadron Collider at the Nuclear Research Centre in Geneva already has the required energy levels. Randall had hoped to be able to provide proof of this, but so far that has not worked out yet.

The hype surrounding string theories has died down somewhat in recent years. Some of its consequences, such as the existence of parallel universes, seem too adventurous. Other attempts at explanation are being sought. Lisa Randall is also at the forefront of this development. She has created several approaches that integrate general relativity into a particle model. None of them has delivered a decisive breakthrough. But then, that is not Randall's expectation. In October 2017, she said in a conversation with science journalist Robert Czepel:

> I am already satisfied in the context of my work if I understand things a little better than was possible before. How is space-time constructed? What does the invisible matter in the universe consist of? These are questions I am interested in. It's about solving problems in small steps – so at least you get an idea of what's behind it all.[4]

Randall is a scientist *par excellence*: she is pursuing a big goal but does not take setbacks personally. She keeps going with new ideas and in this way edges ever closer to a TOE, insight by insight. For example, she has provided significant explanations for the following phenomena, which have caused theoretical physicists a lot of headaches:

- Baryogenesis: In a very early phase of the formation of the universe, baryons were created, i.e., particles with mass that make up ordinary matter. Among others, protons and neutrons are baryons. The Standard Model cannot explain their formation.
- Supersymmetry: This extension is intended to fill some gaps in the Standard Model. It predicts a partner particle for every particle in the Standard Model. Supersymmetry would, among other things, make possible something that is not feasible within the original Standard Model: calculation of the mass of the Higgs boson, which until now could only be determined experimentally.
- Cosmological inflation: Right after the Big Bang, there was a super-fast expansion of the universe, which still puzzles theoretical physicists and cosmologists.

---

[4] Read more at https://science.orf.at/v2/stories/2874349/.

- Dark matter: This phenomenon cannot be explained with the standard model of particle physics either. "Normal" matter can emit electromagnetic radiation, for example with wavelengths in the visible range. Dark matter, on the other hand, is invisible. It is assumed that it cannot emit electromagnetic radiation in any waveband. Only gravity has an effect on it. It is estimated that there is about five times as much dark matter in the universe as normal matter.

* * *

Lisa Randall is not only an outstanding theoretical physicist. She is also something less common in this profession: approachable and understandable. She has mastered the art of presenting the most difficult theoretical trains of thought in such a way that even a layperson can understand them. Randall is known to the public as a best-selling author and contributor to numerous radio and television programmes. Each of her books "*Warped Passages*"[5] and "*Knocking on Heaven's Door*"[6] were included by the New York Times in their list of one hundred most notable books in their years of publication. In her public lectures, Randall is also able to explain complicated physical problems to the audience in a vivid way.

Only very few theoretical physicists have ever managed to make their work known to the public. Their research is too far removed from any generally comprehensible everyday experience. Richard Feynman and Murray Gell-Mann should be mentioned as exceptions here. Lisa Randall belongs to this exclusive circle. From *Newsweek* to *Time Magazine,* important newspapers and magazines have interviewed her and placed her on their lists of important personalities.

Of course, she is also famous in science. Her presence is not only due to the astonishingly large number of her publications—from 1985 to 2022, she published 179 papers, or almost five per year. Some of her papers have been cited nearly 10,000 times, a very high number for papers in her rather arcane field of theoretical physics. The high standing she enjoys in the scientific community is also evidenced by the numerous prizes, awards, honorary memberships, and honorary doctorates with which she has been and continues to be honoured. The only thing missing is the Nobel Prize. So far, however, only very rarely have achievements in abstract theoretical

---

[5] Lisa Randall, *Warped Passages: Unravelling the Mysteries of the Universe's Hidden Dimensions*, Ecco (2006).

[6] Lisa Randall, *Knocking on Heaven's Door: How Physics and Scientific Thinking Illuminate the Universe and the Modern World*. New York: Ecco Press (2011).

physics been honoured with this highest of all scientific awards. Consideration is given almost exclusively to developments that can be directly applied in practice, in keeping with the founder's intentions.

* * *

Lisa Randall is a woman with a wide range of talents, and not only in science. She has also excelled in other domains, including art and philosophy:

- At the invitation of composer Hèctor Parra, she wrote the libretto for his opera *Hypermusic Prologue: A Projective Opera in Seven Planes*. For this work, she took inspiration from her book *Warped Passages*. The opera premiered in 2008 at the Centre Pompidou in Paris, and was subsequently given in Barcelona, Luxembourg, and Brussels. Finally, a special adaptation was chosen as the closing event of the symposium *Universe Resounds: Art & Synaesthesia* at the Guggenheim Museum in New York.
- Randall was curator of the art exhibition *Measure for Measure*, which showed at the Los Angeles Art Galleries.[7] It so happened that Randall became a member of the *American Academy of Arts and Sciences in* 2005, three years before she was accepted as a member of the *National Academy of Sciences*.
- In 2012, she received the *Andrew Gemant Award* from the *American Institute of Physics*, which is given annually for significant contributions to the cultural, artistic, or humanistic dimension of physics.

Lisa Randall is also active in the field of philosophy. She has been a member of the *American Philosophical Society* since 2010. The decisive factor for her membership was her successes as a theoretical physicist. The justification for her membership states:

Lisa Randall's papers with Raman Sundrum on the brane-world with warped extra dimensions are two of the five most highly cited works in high energy theory in the last 20 years. This should not come as a surprise though, as the papers effectively open up new directions in so many different areas of particle theory. Her ideas have shaped the discourse in the field from collider phenomenology to cosmology. An unusually broad and powerful field theorist, Randall has also made important contributions to the theory of supersymmetry breaking and phenomenology, inflation, CP violation, electroweak

---

[7] Read more at https://carpenter.center/program/measure-for-measure.

radiative corrections, the axion, heavy quark physics, and dynamical symmetry breaking.[8]

Incidentally, this shows that the standards and objectives of philosophy in America are clearly different from those in Europe. It should escape nobody that science has a completely different status for American philosophy than for European philosophy?

Lisa Randall is also politically active. She explicitly promotes women physicists and supports the goal of more women getting involved in research and becoming leaders in their field. In her own words:

> If you look through the shelves of science books, you'll find row after row of books written by men. This can be terribly off-putting for women.[9]

Now one last example of Lisa Randall's versatility. She once had an unplanned guest appearance in the American TV series *The Big Bang Theory*. She was visiting some friends at the filming and the film crew immediately seized the opportunity to include her as an extra in the episode—she appears in episode 15 of the third season. Randall got in on the fun and followed the director's instructions very conscientiously: "Just sit there and make yourself invisible!"[10] That she has a sense of humour is shown by her assessment of her performance at the time: "I was really good, by the way!".

Just sit there and make yourself invisible … Since ancient times, women have been expected to comply with this injunction. For a long time, nobody wanted to admit that they could be just as hungry for education, just as curious, and just as creative as men. Only a few women succeeded in gaining a place in science. In almost every case, they only had a real chance if they were born into educated and cosmopolitan families. The potential of countless others was wasted.

Today, it has become easier for women to make a career in science. Even though beliefs such as "girls become hairdressers, not scientists" still circulate and unfair assessments fuel self-doubt, women have fought for their place in science. Some universities even consciously strive to increase the proportion of women and advertise jobs explicitly for women. Resistance is already making itself felt from some male colleagues who feel left out.

---

[8] On the website of the *American Philosophical Society*: https://search.amphilsoc.org/memhist/search?creator=Lisa+Randall&title=&subject=&subdiv=&mem=&year=&year-max=&dead=&keyword=&smode=advanced.

[9] Read more at www.wisefamousquotes.com/lisa-randall-quotes/.

[10] Interview with Robert Czepel on 27.10.2017, available at https://science.orf.at/v2/stories/2874349/.

Just sit there and make yourself invisible ... it's a good thing Lisa Randall has only followed this stage direction once in her life. She spends the rest of her time researching how gravity works at CERN's Large Hadron Collider in Geneva, uncovering the mystery of dark matter, and integrating the Standard Model of particle physics and general relativity into a Theory of Everything.

18

Maryam Mirzakhani (1977–2017): The First Female Recipient of the Fields Medal

Since 1936, the Fields Medal has been awarded as the highest scientific distinction at the International Congress of Mathematicians, which takes place every four years. The prize is awarded to two or more mathematicians who must not be older than 40. Until 2014, the total of 52 laureates of this unofficial Nobel Prize in mathematics were all men. It was not until that year, 78 years after the prize was founded, that it was awarded to a woman for the first time: to the Iranian Maryam Mirzakhani. Just like Emmy Noether in the 1930s, Mirzakhani had created astonishingly abstract structures that would be incomprehensible to non-mathematicians, but have nevertheless found their way into our everyday lives.

The mathematical topics Mirzakhani dealt with were essentially the following:

1. Hyperbolic geometry: When we imagine triangles, straight lines, angles, etc., we move in Euclidean space, named after the ancient Greek mathematician Euclid. Hyperbolic geometry differs from Euclidean geometry in one single criterion: parallelism. In a Euclidean plane, for every given straight line and a point that does not lie on this straight line, there is always exactly one other straight line that passes through the given point and runs parallel to the first straight line. The two parallels are always the same distance apart and never touch. In hyperbolic geometry, on the other hand, there are at least *two* straight lines that go through the point and are

© The Author(s), under exclusive license to Springer Nature
Switzerland AG 2023
L. Jaeger, *Women of Genius in Science*,
https://doi.org/10.1007/978-3-031-23926-7_18

parallel to the original straight line. Hyperbolic spaces therefore massively contradict our experience and imagination.

2. Symplectic geometry: This is a non-Euclidean geometry that is even more abstract and complex than hyperbolic geometry. It is a branch of differential geometry, which deals with curved surfaces that have neither holes nor sharp edges. Mathematicians call such surfaces differentiable. A differentiable plane is said to be symplectic if—to put it simply—a*b = b*a does not apply to it, but a*b = −b*a. This abstract geometry is also counterintuitive, but it has great significance for theoretical physics, and especially quantum theory.

3. The Teichmüller and ergodic theories are many times more abstract and complex than the geometries mentioned above. The former deals with so-called compact Riemann surfaces, while the latter is a combination of measure theory, a generalisation of the intuitive notions of lengths, areas, and volumes, stochastics which is used to calculate probability distributions, and the theory of dynamic systems. In reference to such highly abstract systems, it is said to have been Richard Feynman who once said, "*Shut up and calculate!*" In other words, don't even try to imagine why something in abstract mathematics is the way it is. Convince yourself that the calculation steps are logical in themselves and just get on with it!

If one imagines abstract mathematics as an extension to the house of classical mathematics, Emmy Noether had fully pushed open the door to this extension in the first decades of the twentieth century and also discovered that it was very much larger than the original building. Then, one hundred years later, the mathematician Maryam Mirzakhani learnt to move very skilfully around the floors of this annex, always looking for new corridors, opening up the rooms along them, and finding connecting doors.

* * *

Maryam Mirzakhani was born in Tehran, the capital of Iran, on 12 May 1977, the third of four children. Her father Ahmad Mirzakhani was an electrical engineer, her mother Zahra a housewife. As a child, Maryam was lively, curious, and smart. She loved reading and learning. Sometimes she would persuade her brother Arash, who was six years older, to tell her about what he had learned at school. Once he challenged her to find the sum of all the numbers from 1 to 100, and then told her about the ingenious solution that the mathematician Carl Friedrich Gauss had found as a child. Gauss had realised that the numbers in the series from 1 to 100 could be paired in such a way as always to add up to 101: 100 + 1 = 101; 99 + 2 = 101; 98 + 3

= 101 and so on. So, he only had to calculate 50 times 101 and he would immediately come up with the solution: 5050. Maryam was thrilled by this trick:

> That was the first time I enjoyed a beautiful solution, though I couldn't find it myself.[1]

This anecdote gives us a hint of another outstanding quality that, along with intelligence and curiosity, was characteristic of Miryam Mirzakhani's personality: her positive nature. Instead of being angry that she had not been able to find an answer, or even envious of someone who had succeeded in doing so, the aesthetics of the solution filled her with happiness.

This trait was supported by her loving parents, who set the example to Maryam and her siblings that staying true to one's values is more important than worrying about money and success. Among the qualities highly respected in the parental home were modesty and helpfulness. Maryam's sister Leila later described a happy childhood with strong-minded parents who gave their children love, serenity, peace, and respect and expected them to make the most of their lives: "It was a place I would call heaven."[2]

In stark contrast to the security within the family were the political events in Mirzakhani's home country. In 1978, one year after Miryam's birth, unrest and mass protests began in Iran. Within a short time, this unrest had escalated into the Iranian Revolution. The ousting of the Shah in 1979 was followed by the return of the Ayatollah Khomeini from exile and the proclamation of the Islamic Republic of Iran, which remains an orthodox theocracy to this day. Miryam's parents did everything they could to give their children a carefree childhood, despite what was going on around them. And when Iran was at war with Iraq, from 1980 to 1988, with disastrous losses incurred on both sides, they also managed to shelter the children from any consequences. When Miryam was eleven, the age when her parents could no longer protect her from her own experience of the political situation, times became calmer again. Later, she would say that she was lucky to belong to a generation that was no longer directly affected by the riots and hardships that accompanied the establishment of the Republic of Iran.

And Miryam Mirzakhanis was also lucky in another respect: unlike today's Taliban rule in Afghanistan, the government of the Ayatollah Khomeini did not exclude girls from higher education. Maryam attended Tehran's

---

[1] "Interview with Research Fellow Maryam Mirzakhani", Clay Mathematics Institute Annual Report (2008), pp. 11–13; available at www.claymath.org/library/annual_report/ar2008/08AnnualReport.pdf.
[2] Read more at www.egmo2018.org/blog/wimbs-maryam-mirzakhani-part3/.

Farzanegan Girls' School, which offered special middle and high school classes for exceptionally talented students. Before Maryam discovered her talent for mathematics here, another career aspiration was at the forefront of her mind:

> As a kid, I dreamt of becoming a writer. (…) I never thought I would pursue mathematics before my last year in high school.[3]

Maryam Mirzakhani did not believe that she is mathematically gifted. This was probably also due to a teacher who once told her that she had absolutely no talent in mathematics—a remark that many a young girl must have heard and taken to heart. But Maryam's enthusiasm for mathematics would soon reach a turning point. It happened in a mathematics summer workshop for pupils from the age of 16 with exceptional mathematical talent. Maryam was allowed to take part in it, even though she was only 15 years old. A problem posed in the workshop by Professor Ebad Mahmoodian concerned the decomposition of graphs. Most readers will remember the lines they were asked to draw in a two-dimensional coordinate system—straight lines, parabolas, asymptotes, and so on. In mathematics, however, a graph is defined much more generally as a set of nodes ("points") connected by edges ("links"). The students attending Mahmoodian's summer camp were long past the stage where the usual graphs from normal school lessons might interest them. The task given to them was to decompose a certain type of graph into what were called 5-cyclic graphs.

As a first step, Mahmoodian suggested that students construct an example of a graph of this particular type. Since this was already a very challenging task for 16-year-old students, he offered a prize of one dollar for each example they were able to find. After a short time, Maryam discovered a whole family of graphs with the required property; a family with an infinite number of members—the prize money was never paid out.[4] Impressed, Mahmoodian invited Maryam to continue working on the problem with him at Tehran's Sharif University. In 1995, he published the solution to the problem with the now 17-year-old Maryam Mirzakhani.[5]

Maryam was gradually drawn further and further into mathematics. Just as in sport, there are also olympiads in mathematics. Since 1959, participating

[3] "Interview with Research Fellow Maryam Mirzakhani", Clay Mathematics Institute Annual Report (2008), pp. 11–13; available at www.claymath.org/library/annual_report/ar2008/08AnnualReport.pdf.

[4] Statement by Ramin Takloo-Bighash in: Hélène Barcelo and Stephen Kennedy: Maryam Mirzakhani (1977–2017), Notices of the AMS, November 2018; available at: www.ams.org/journals/notices/201810/rnoti-p1221.pdf.

[5] Ebad Mahmoodian and Maryam Mirzakhani, *Decomposition of Complete Tripartite Graphs into 5-Cycles Mathematica Applications*, 324, Dordrecht: Kluwer Academic Publications (1995).

countries have been able to send up to six school pupils to the International Mathematical Olympiad IMO every year. These young people are selected in preceding national olympiads. Mahmoodian persuaded the relevant officials to admit Maryam Mirzakhani to the National Mathematical Olympiad. Never before had a girl taken part in this selection process in Iran. Moreover, Maryam was still in the tenth grade and not in the upper school, as the entry requirements stipulated. Once again, Mirzakhani was a year too young, and again an exception was made for her. In the national event she won a gold medal and was included in the squad to represent Iran at the IMO held in Hong Kong in July 1994. In Hong Kong, she scored 41 out of 42, a rare achievement, and won a gold medal. The following year she was back again and won her second international gold medal in Toronto with a full score. At that olympiad, 73 countries were represented, and among the total of 439 young people, only 27 were girls.

A good friend of Maryam's from Farzanegan Girls' School, Roya Beheshti, also took part in the Hong Kong olympiad, where she won a silver medal with 35 out of 42 points. After graduating from school, Mirzakhani and Beheshti began studying at Tehran's Sharif University, which was founded in 1344 as an elite university. Noohi Behrang, a former instructor for the International Mathematical Olympiad and now a lecturer in the Department of Mathematics at *Queen Mary University of London*, said:

> She told me that she wanted to continue math at the university because she found mathematicians nice people. She loved math but at the same time she would see the human side of it, the kindness and the niceness of the people there was important to her, and she was indeed an example of a kind and nice person.[6]

On 17 March 1998, the story of Miryam Mirzakhani and Roya Beheshti almost came to an end. The two friends attended a conference in the Iranian city of Ahvaz, 800 kms away, and while they were travelling back to the capital in a bus together with other highly gifted students, the bus crashes into a ravine. Seven students and the two bus drivers lost their lives. Mirzakhani and Beheshti were lucky and narrowly escaped death. It was also lucky for mathematics. A year later, the two published the book *Elementary Number Theory, Challenging Problems.*[7] Today, Beheshti teaches at the renowned MIT.

* * *

---

[6] www.egmo2018.org/blog/wimbs-maryam-mirzakhani-part4/.

[7] Maryam Mirzakhani, Roya Beheshti, Elementary Number Theory, Challenging Problems, Fatemi Publishers, Iran (1999; the book is in Persian).

At first, mathematics was still a game for Maryam, but thanks to the mathematicians she met, that game became a passion:

> As a teenager, I enjoyed the challenge. But most importantly, I met many inspiring mathematicians and friends at Sharif University. The more I spent time on mathematics, the more excited I became.[8]

Even before Mirzakhani graduated from Sharif University with a bachelor's degree in 1999, she received an award from the *American Mathematical Society* for her work on developing a simple proof of a mathematical theorem dating back to 1916.[9] Not many bachelor students can boast such success.

Mirzakhani now moved to Harvard University in the United States to do a PhD on Riemann surfaces; her supervisor was one of the 1998 Fields Medal winners, Curtis McMullen. One should not be deceived by the word "surface," for these are not two-dimensional, but abstract multi-dimensional constructions, which in the simplest case have *locally* the structure of the complex number plane. Simplified, Riemannian surfaces can be imagined as objects containing several holes. In highly simplified terms, Mirkazhani was trying to work out how many different windings of a ribbon of a certain length could be wrapped around a Riemann surface—for example, a surface a bit like two donuts pressed together to form a figure of eight—without the ribbon crossing itself. Because she was not actually dealing with two donuts but with complex number planes, this task was of course much more abstract. Maryam Mirzakhani realised that the problem could be reversed: instead of fixing the figure and counting the number of windings, she calculated the average of all the numbers corresponding to points in a moduli space of Riemann surfaces. A moduli space is a set of points that represents one of the shapes that a surface can take in three dimensions. To calculate this average, she had to calculate the size, or "volume," of certain subspaces of Riemann surfaces. Mirzakhani found a clever formula for the volumes and thus solved the problem.

In 2004, she completed her doctorate with this work.[10] The 27-year-old then turned down the junior fellowship offered to her by Harvard and became a research fellow at the *Clay Mathematics Institute* in Princeton University in New Jersey, where she also began to teach. In 2008, at only

---

[8] Interview in *The Guardian*, 13 August 2014.

[9] Maryam Mirzakhani, A Simple Proof of a Theorem of Schur, *The American Mathematical Monthly*, 105, 3 (1998).

[10] Miryam Mirzankhani, *Simple Geodesics on Hyperbolic Surfaces and the Volume of the Moduli Space of Curves*. Available at https://www.math.stonybrook.edu/~mlyubich/Archive/Geometry/Teichmuller%20Space/Mirz3.pdf.

31, she became a full professor at Stanford University in California. In under ten years, a graduate in Iran had become one of the world's most important mathematicians. In her early 30s, Maryam Mirzakhani was a leader in the fields of hyperbolic geometry, topology, and dynamics.

Following her doctoral thesis, she published further contributions to the theory of Riemann surfaces. She studied geodesics, which are curves that connect two points by the shortest possible path. On two-dimensional planes, the matter is simple: here, the shortest connection is always a straight line. Even in a three-dimensional space, geodesics can be counterintuitive. If, for example, two points are to be connected in a shoe box, one of which is in the middle of a side edge on the box floor and the other on the opposite edge of the upper cover, then the geodesic does not run across the box floor to the opposite wall and then up the wall to the second point, but across the box floor, one of the sides, and the box upper cover. Geodesics are also important, for example, in the multi-dimensional spaces that appear in Einstein's theory of gravity. We are familiar with the three-dimensional image of a kind of rubber mat in which the Earth's sphere lies in a deep hollow caused by its mass. If we want to calculate the effects of gravity on the way an object moves, we need to find the shortest paths on this strangely deformed rubber mat (which in reality is of course not three-dimensional but multi-dimensional).

Mirzakhani's next project was somewhat related to geodesics: she investigated the dynamics of a point mass (which can also be imagined as a billiard ball) moving inside a polygon, i.e., a regular plane shape with n sides. As in billiards, the point moves in a straight line until it hits the edge of the polygon; then it bounces back at the same angle at which it hit. Mathematicians have asked several questions about this process. One of them is: will the path of the point within the given polygon ever repeat itself? And if so, how many such paths are there, and what do they look like in detail? Or will the motion eventually touch every point on the polygon? The answer to the question of whether repeating paths can be found in *all* polygons still remains to be answered. But along the way, Mirzakhani and her colleague Alex Eskin developed a method that embeds the space of the polygonal billiard table in a space with more dimensions. The points are also lifted into a higher dimension, where they become surfaces that are locally either flat or conical.

With this trick, Mirzakhani and Eskin were able to prove a certain theorem about a group of symmetrical geometric objects that had long been a headache for mathematicians. Unlike most mathematical proofs, which are often only a few pages long, their proof extends to no fewer than 200 pages.[11]

[11] Alex Eskin and Marjam Mirzakhani, *Invariant and Stationary Measures for the SL(2,R) Action on Moduli Space*. Available at https://arxiv.org/abs/1302.3320 (2013).

The theorem proven by Mirzakhani and Eskin is called a "magic wand" by mathematicians because it enables the solution of many other previously unsolvable mathematical problems. For example, it can be used to model the dispersion of gases. To the astonishment of most mathematicians, Mirzakhani's trick, which introduces and uses a higher number of dimensions, does not lead to a more complex behaviour of the equations. They would have expected a fractal behaviour of the equations and a slide into chaotic processes. Instead, the systems stabilise.

* * *

Mirzakhani has received many awards throughout her career, here are just a few examples:

- In 2009, her dissertation was awarded a prize by the *American Mathematical Society*.
- In 2013, she received the coveted *Satter Prize,* which is awarded every two years to outstanding female mathematicians.
- In 2014, she was awarded the *Clay Research Award*, an annual award which honours outstanding mathematical achievements.

On 13 August 2014 came the highlight of Maryam Mirzakhani's career: she was awarded the Fields Medal. In his laudation, the American mathematician Jordan Ellenberg explained Mirzakhani's research results as follows:

[Her] work expertly blends dynamics with geometry. Among other things, she studies billiards. But now, in a move very characteristic of modern mathematics, it gets kind of meta: She considers not just one billiard table, but the universe of all possible billiard tables. And the kind of dynamics she studies doesn't directly concern the motion of the billiards on the table, but instead a transformation of the billiard table itself, which is changing its shape in a rule-governed way; if you like, the table itself moves like a strange planet around the universe of all possible tables [...] This isn't the kind of thing you do to win at pool, but it's the kind of thing you do to win a Fields Medal.[12]

Even at this high point of her career, Maryam Mirzakhani remained modest, saying at the ceremony in her department that there were so many other people of high merit who should have won the coveted award instead

---

[12] Jordan Ellenberg, *Math Is Getting Dynamic*, 14 August 2014; available at www.slate.com/art icles/life/do_the_math/2014/08/maryam_mirzakhani_fields_medal_first_woman_to_win_math_s_big gest_prize_works.html

of her. Hardly anyone shared her opinion, however. It was clear among her colleagues that she undoubtedly deserved this award.

The Fields Medal also helped Mirzakhani to achieve great popularity in her home country. Iranian President Hassan Rouhani proudly congratulated her on winning the most important mathematics prize in the world.

Mirzakhani could easily have become a media star, but she tended to avoid publicity. She never pushed to speak out in front of the world on political or social issues. So, all the more remarkable was her speech to the *American Mathematical Society in* 2013, in which she noted that the situation for women in mathematics is still far from ideal:

> The social barriers for girls who are interested in mathematical sciences might not be lower now than they were when I grew up. And balancing career and family remains a big challenge. It makes most women face difficult decisions which usually compromise their work.[13]

* * *

Parallel to her highly successful career as a mathematician, Maryam Mirzakhani managed to lead a fulfilling family life. In 2008, she married the Czech Jan Vondrák, who works in theoretical computer science and applied mathematics and is now an associate professor at Stanford University. In 2011, their daughter Anahita was born.

But fate also had bad things in store for Maryam Mirzakhani. When she received the Fields Medal in 2014, she was already suffering from breast cancer. The journey to the award ceremony in Seoul was a huge strain for her, leaving her exhausted. Ingrid Daubechies, a mathematics professor at Princeton with many awards herself, showed great foresight when she instructed others to keep overly intrusive congratulators and members of the press at a distance from the laureate. While the celebrations were still going on, Mirzakhani had to leave the event. But she was able to receive the Fields Medal in person, to thunderous applause.

Even as cancer returned in 2017, Mirzakhani always remained positive in her outlook. As she pointed out, she had been born into a loving family and had a good mind, and not everyone could say that.[14]

---

[13] Maryam Mirzakhani, in a 2013 talk given at the *American Mathematical Society* after receiving the Ruth Lyttl Satter Prize; http://www.ams.org/notices/201304/rnoti-p490.pdfi.

[14] https://news.stanford.edu/2017/10/23/colleagues-friends-family-gather-remember-stanford-profes sor-maryam-mirzakhani/.

On 14 July 2017, Miryam Mirzakhani died at Stanford at the age of only 40. She knew what was in store for her; her husband Jan Vondrák said at the funeral service held in her honour at the Stanford University auditorium:

> She said don't be too quick to cry for me. There's a lot of trouble in the world. Cry for those who are close to you and who you can help.[15]

Eleny Ionel, a Romanian-born professor and chair of the mathematics department at Stanford, was in a far more emotional mood:

> It is still very hard to imagine that someone of her extraordinary energy, determination and brilliance could be taken away from us at such a young age. Her passion for mathematics has touched so many lives and will continue to be an inspiration for many more.[16]

Iranian newspapers broke the taboo of publishing pictures of women without head coverings. Since Mirzakhani had stopped wearing a headscarf soon after she left for the USA, there were no recent photos of her with the covering prescribed in Iran. Editors had to resort to pictures showing her without the hijab. Even one of the pictures posted on Twitter by Iranian President Hassan Rohani shows Mirzakhani without a headscarf. His statement bears witness to the importance of Mirzakhani's achievement in a country that today has few friends in the world community because of its aspirations to build nuclear bombs:

> The grievous passing of Maryam Mirzakhani, the eminent Iranian and world-renowned mathematician, is very much heartrending.[17]

Maryam Mirzakhani enriched mathematics and opened up whole new territories for research. And although she had always remained apolitical during her lifetime, her death set other things in motion:

- In the wake of Mirzakhani's death, the Iranian parliament passed a law that allows children of Iranian mothers who, like Mirzakhani, are married to foreigners to take Iranian citizenship. Previously, children from such marriages had great difficulty entering Iran.

---

[15] Ibid.

[16] Ibid.

[17] Maryam Mirzakhani: Iranian Newspapers Break Hijab Taboo in Tributes, *The Guardian*, 16 July 2017; available at: www.theguardian.com/world/2017/jul/16/maryam-mirzakhani-iranian-newspapers-break-hijab-taboo-in-tributes.

- In 2018, delegates at an international congress of women mathematicians in Rio de Janeiro decided to establish 12 May, Maryam Mirzakhani's birthday, as an annual global day of celebration for women in mathematics.
- Since 2021, the *Maryam Mirzakhani New Frontiers Prize has been* awarded every year to outstanding young female mathematicians.
- In 2022, the 37-year-old Ukrainian mathematician Maryna Viazovska, who works at the Polytechnic University in Lausanne, became the second woman to win the Fields Medal.

Epilogue

Many scholars have never heard the names of Émilie du Châtelet, Laura Bassi, or Sofia Kovalevskaya. Even the impact of women in the twentieth century remains unknown to many. The exception here is perhaps Marie Curie with her two Nobel Prizes, but who has heard of Emmy Noether, Chien-Shiung Wu, Rosalind Franklin, Jocelyn Bell Burnell, or Grete Hermann?

And there are many other women of scientific genius who are not explicitly included in the book. Indeed, it would be impossible to include all of them in a book like this. But let us mention four great names and share some brief information about each.

1. Maria Goeppert Mayer

    Maria Goeppert Mayer (1906–1972) was a German-born American theoretical physicist. She was only the second woman to win a Nobel Prize in physics, awarded exactly 50 years after Marie Curie, for her nuclear shell model of the atomic nucleus.

2. Barbara McClintock

    Barbara McClintock (1902–1992) was an American scientist and early cytogeneticist (a branch of genetics that analyses chromosomes primarily with a light microscope) who was awarded the Nobel Prize in Physiology/Medicine in 1983. To date, she is the only woman to have received an undivided Nobel Prize in Physiology/Medicine.

3. Henrietta Swan Leavitt

    Henrietta Swan Leavitt (1868–1921) was a successful American astronomer who developed the first "standard candle" that could be used

© The Editor(s) (if applicable) and The Author(s), under exclusive license to Springer Nature Switzerland AG 2023
L. Jaeger, *Women of Genius in Science*,
https://doi.org/10.1007/978-3-031-23926-7

to establish the distance to remote galaxies. Given the absolute luminosity of such an object, a function of periodic variations in its apparent brightness, its distance can then be determined.

4. Vera Rubin

Vera Rubin (1928–2016) was an American astronomer who pioneered work on the rotation rates of galaxies. She discovered a discrepancy between the predicted and observed (angular) motion of galaxies. Rubin studied galactic rotation curves in detail and her work provided the first evidence for the existence of dark matter (matter that cannot be detected by electrodynamic waves). Her results were confirmed in detail in later decades.

The position of women in the sciences has improved greatly, at least in Western countries. The fact that more and more women are reaching the top of science and mathematics is also shown by the award of the Fields Medal 2022 to Maryna Viazovska, who is 37 years old and only the second woman after Maryam Mirzakhani to receive this highest award. Nevertheless, we must continue to increase the number of women at the top. The total number of women with a Nobel Prize in physics, which has been around for over 120 years, is still just four, so apart from Marie Curie, there are only three others. In chemistry and medicine/biology there are seven, in physiology/medicine twelve.

Meanwhile, it is pretty well proven that the aptitude of women for the natural sciences (as based on intelligence quotients) is the same as for men. What is probably true is that there is a greater dispersion among men, i.e., more outliers, upwards as well as downwards, i.e., more superintelligent people (with an IQ above 130), but also more mentally limited people (with an IQ below 70). There is also a genetic reason for this greater dispersion: men have an X and a Y chromosome and a large part of the hereditary dispositions that help determine mental abilities are located on the X chromosome. To simplify somewhat, if one or more of the relevant genes mutates, the corresponding man becomes either particularly clever or particularly stupid. The same changes can, of course, also occur in women, but in men, a change on just one X chromosome has the full effect, whereas women still have a second chance, which can compensate for the effect on the first X chromosome and thus mitigate the mutation.

Of course, IQ is not the only measure of potential scientific success. Also relevant are creativity, the ability to work in a team, perseverance, and many other criteria where there is no difference between men and women.

Bibliography

Abaelard, Peter, *Sic et non (Yes and No)*, Primary Source Edition (2014).

Abtei St. Hildegard (Hg.), Hildegard von Bingen, *Causae et Curae*, newly translated, Wentworth Press (2019).

Asmus, Johann Rudolf (Hg.), *Das Leben des Philosophen Isidoros von Damaskios aus Damaskos*, Meiner, Leipzig (1911).

Bodanis, David, *Einstein's Greatest Mistake: A Biography*, Mariner Books (2017).

Bölling, Reinhard (Hg.), *Briefwechsel zwischen Karl Weierstraß und Sofia Kowalewskaja*, Akademie Verlag Berlin (1993).

Bowden, Bertram V. (Hg.), *Faster Than Thought—A Symposium on Digital Computing Machines*, Pitman (1953).

Ceranski, Beate, *Und sie fürchtet sich vor niemandem. Die Physikerin Laura Bassi (1711–1778)*, Campus-Verlag, Frankfurt a. M. (1996).

Charles, Robert Henry, *The Chronicle of John, Bishop of Nikiu*, London (1916).

Chiang, Tsai-Chien, *Madame Chien-Shiung Wu: The First Lady of Physics Research*, World Scientific Publishing Company (2013).

Crull, Elise, Bacciagaluppi, Guido (Hg.), *Grete Hermann—Between Physics and Philosophy*, Springer (2016).

Curie, Marie, *Traité de Radioactivité*, Gauthier-Villars, Paris (1910).

Curie, Marie, *La Radiologie et la Guerre*, Félix Alcan, Paris (1921).

Del Centina, Andrea, Unpublished Manuscripts of Sophie Germain and a Revaluation of Her Work on Fermat's Last Theorem, *Archive for History of Exact Sciences*, 62 (4) (2008).

Einstein, Albert, *Collected Papers*, 9B, n. 548, Princeton University Press (1987).

Fell, James, Elementare Beweise des großen Fermat'schen Satzes für einige besondere Fälle, *Deutsche Mathematik*. 7 (1943).

© The Editor(s) (if applicable) and The Author(s), under exclusive
license to Springer Nature Switzerland AG 2023
L. Jaeger, *Women of Genius in Science*,
https://doi.org/10.1007/978-3-031-23926-7

Findlen, Paula, *Studies on the History of Society and Culture*, University of California Press, Vol. 84, Book 3 (1993).

Goodall, Jane, *Through a Window: My Thirty Years with the Chimpanzees of Gombe* (1986); today under Mariner Books (2010).

Goodall, Jane, *Through a Window*, Mariner Books (1990).

Grützmacher, Georg, *Synesios von Cyrene: ein Charakterbild aus dem Untergang des Hellenentums* (German Edition), A. Deichert'sche Verlagsbuchhandlung, Leipzig (1913).

Hardy, Anne, Sexl, Lore, *Lise Meitner* (German Edition), rororo (2002).

Hagengruber, Ruth, Hecht, Hartmut, *Émilie du Châtelet und die deutsche Aufklärung*, Springer Verlag (2019).

Heisenberg, Werner, *Physics and Beyond: Encounters and Conversations*, Harper & Row (1972).

Hermann, Grete, *Die naturphilosophischen Grundlagen der Quantenmechanik*, Abhandlungen der Fries'schen Schule (ASFNF), Bd. 6, Heft 2 (1935).

Herrmann, Kay (Hg.), *Grete Hermann: Philosophie – Mathematik – Quantenmechanik*, Springer, Wiesbaden (2019).

Herschel, Caroline, *Catalogue of Stars Taken from Mr. Flamsteed's Observations Contained in the Second Volume of the* Historia cœlestis, *and not Inserted in the British Catalogue. With an Index, to Point Out Every Observation in That Volume Belonging to the Stars of the British Catalogue. To Which Is Added, a Collection of Errata That Should Be Noticed in the Same Volume*, Published by Order, and at the Expense, of the Royal Society, London (1798).

Imhof, Agnes, *Die geniale Rebellin: Ada Lovelace – Sie stürzte sich ins Leben und revolutionierte die Mathematik*, Piper Taschenbuch (2022).

Isaacson, Walter, *Einstein: His Life and Universe,* Simon & Schuster, New York (2008).

Jacobson, Nathan (Hg.), *Emmy Noether – Gesammelte Abhandlungen – Collected Papers*, Springer, Berlin, Heidelberg (1983).

Jaeger, Lars, *Die Naturwissenschaften – Ein Biographie*, Springer, Heidelberg (2015).

Jaeger, Lars, *Wissenschaft und Spiritualität – Universum, Leben, Geist, Zwei Wege zu den großen Geheimnissen*, Springer, Heidelberg (2016).

Jaeger, Lars, *Supermacht Wissenschaft – Unsere Zukunft zwischen Himmel und Hölle*, Gütersloher Verlagshaus, Gütersloh (2017).

Jaeger, Lars, *The Second Quantum Revolution: From Entanglement to Quantum Computing and Other Super-Technologies*, Springer, Heidelberg (2018).

Jaeger, Lars, *Mehr Zukunft wagen – Wie wir alle vom Fortschritt profitieren*, Gütersloher Verlagshaus, Gütersloh (2019).

Jaeger, Lars, *Sternstunden der Wissenschaften - Eine Erfolgsgeschichte des Denkens*, Südverlag, Konstanz (2020).

Jaeger, Lars, *Ways Out of the Climate Catastrophe: Ingredients for a Sustainable Energy and Climate Policy*, Springer, Heidelberg (2021).

Jaeger, Lars, *Emmy Noether – Ihr steiniger Weg an die Weltspitze der Mathematik*, Südverlag, Konstanz (2022).

Jaeger, Lars, *The Stumbling Progress of 20th Century Science: How Crises and Great Minds Have Shaped Our Modern World*, Springer, Berlin (2022).

Kerner, Charlotte, *Lise, Atomphysikerin – Die Lebensgeschichte der Lise Meitner*, Beltz und Gelberg (1996).

Koreuber, Mechthild, *Emmy Noether, die Noether-Schule und die moderne Algebra: Zur Geschichte einer kulturellen Bewegung*, Springer Spektrum, Berlin, Heidelberg (2015).

Kosmann-Schwarzbach, Yvette, Schwarzbach, Bertram, *The Noether Theorems. Invariance and Conservation Laws in the Twentieth Century*, Springer, New York (2011).

Kowalewskaja, Sofia, *A Russian Childhood*, Springer (1978; original in 1890); accessible: Projekt-Gutenberg.org.

Kowalewskaja, Sofia, *Autobiographische Skizze*, Deutsche Rundschau, 108 (1901).

Lovelace, Ada, *Notes to a "Sketch of the Analytical Engine Invented by Charles Babbage, by L.F. Menabrea,"* Scientific Memoirs, Band 3, London (1843).

Legendre, Adrien-Marie, *Recherches sur quelques objets d'analyse indéterminée et particulièrement sur le théorème de Fermat*, in *Mémoires de l'Académie royale des sciences de l'Institut de France*, Vol. 6 (1823).

Kagele, Hanna, *Sophie Germain, The Princess of Mathematics and Fermat's Last Theorem.* https://www.gcsu.edu/sites/files/page-assets/node-808/attachments/kagele.pdf.

Mackinnon, Nick, Sophie Germain, or "Was Gauss a Feminist?", *The Mathematical Gazette*, 74 (470) (1990) S. 346–351.

Mädler, Johann Heinrich, *Geschichte der Himmelskunde* (1873).

Mayrn, Heinrich, Pinzger, Martin (Hg.), *Informatik 2016, Lecture Notes in Informatics (LNI)*, Gesellschaft für Informatik, Bonn (2016).

McGrayne, Sharon Bensch, *Nobel Prize Women in Science: Their Lives, Struggles, and Momentous Discoveries*, Carol Publishing Group, Secaucus (1993).

Maddox, Brenda, *Rosalind Franklin: The Dark Lady of DNA*, HarperCollins, London (2002).

Menabrea, Luigi, *Notions sur la machine analytique de M. Charles Babbage*, Bibliothèque universelle de Genève, nouvelle série 41 (1842).

Menabrea, Frederico, Lovelace, Ada, *Sketch of the Analytical Engine Invented by Charles Babbage: Translation and Notes by Ada Lovelace*, Independently published (2020).

Mirzakhani, Maryam, Beheshti, Roya, *Elementary Number Theory, Challenging Problems*, Fatemi Publishers, Iran (1999; das Buch ist in persischer Sprache verfasst).

Novak, Ralph, *Christianity and the Roman Empire*, Trinity Press International Harrisburg, PA (2001).

Osen, Lynn, *Women in Mathematics*, The MIT Press, Cambridge, London (1974).

Piani, Domenico, *Catalogo dei Lavori dell'Antica Accademia, raccolti sotto i singoli autori* (1852). A new version in: Elena, Alberto, *In lode della filosofessa di Bologna: An Introduction to Laura Bassi*, Isis, Vol. 82, Number 3 (1991).

Popp, Karl, Stein, Erwin (Hg), *Gottfried Wilhelm Leibniz. Das Wirken des großen Universalgelehrten als Philosoph, Mathematiker, Physiker, Techniker*. Schlütersche, Hannover (2000).

Quinn, Susan, *Marie Curie: A life*, Addison-Wesley (1996).

Rademacher, André (Hg.), *Hildegard von Bingen, Der Mensch in der Verantwortung – Das Buch der Lebensverdienste (Liber Vitae Meritorum)*, Otto Müller Verlag, Salzburg (1986).

Randall, Lisa, *Warped Passages: Unraveling the Mysteries of the Universe's Hidden Dimensions*, Ecco Press (2006).

Randall, Lisa, *Knocking on Heaven's Door: How Physics and Scientific Thinking Illuminate the Universe and the Modern World*, Ecco Press, New York (2011).

Rowe David, Koreuber, Methild, *Proving It Her Way. Emmy Noether, a Life in Mathematics*, Springer (2020).

Smith, Martha, Brewer, James (Hg.), *Emmy Noether: A Tribute to Her Life and Work*, Marcel Dekker, New York (1981).

Somerville, Mary, *The Magnetic Properties of the Violet Rays of the Solar Spectrum*, Proceedings of the Royal Society (1826).

Srinivasan, Bhama, Sally, Judith (Hg.), *Emmy Noether in Bryn Mawr – Proceeding of a Symposium*, sponsored by the Association for Women in Mathematics in Honor of Emmy Noether's 100th birthday, Springer, Berlin (1983).

Tannery, Paul (Hrsg.), *Œuvres de Fermat. Tome premier*. Gauthier-Villars, Paris (1891).

Vogt, Joseph, *Begegnung mit Synesios, dem Philosophen, Priester und Feldherrn. Gesammelte Beiträge*, Darmstadt (1985).

Vögtle, Fritz, Ksoll, Peter, *Marie Curie*, Rowohlt (2018).

Voltaire, *Preface About Marquise du Châtelet*, in I. Newton, *Principes mathématiques de la philosophie naturelle*. New edition in the French version (1990).

von Bingen, Hildegard, *Causae et Curae*, translated by Manfred Pawlik and Patrick Madigan, edited by Mary Palmquist and John Kulas, Liturgical Press, Inc. (1994).

von Bingen, Hildegard, *Im Feuer der Taube: die Briefe*, Übersetzt und herausgegeben Walburga Storch, Pattloch, Augsburg (1997).

von Bingen, Hildegard, *Physica*, translated into English by Priscilla Throop, Healing Arts Press Rochester (1998).

von Bingen, Hildegard, *Ursachen und Behandlung der Krankheiten*, Lempertz Klassiker, Königswinter (2013).

Von Neumann, John, *Mathematical Foundations of Quantum Mechanics*, originally 1932, new edition: Princeton University Press (2018).

Watts, Edward, *Hypatia—The Life and Legend of an Ancient philosopher*, Oxford University Press (2017).

Watson, James, *The Double Helix: A Personal Account of the Discovery of the Structure of DNA*, Signet (1998).

Zinssers, Judith (Herausgeberin), *Selected Philosophical and Scientific Writings*, University of Chicago Press (2009).

Zitelmann, Arnulf, *Hypatia*, Beltz & Gelberg Weinheim, Basel (1988).